Idealization in Epistemology

Idealization in Epistemology

A Modest Modeling Approach

DANIEL GRECO

OXFORD
UNIVERSITY PRESS

OXFORD
UNIVERSITY PRESS

Great Clarendon Street, Oxford, OX2 6DP,
United Kingdom

Oxford University Press is a department of the University of Oxford.
It furthers the University's objective of excellence in research, scholarship,
and education by publishing worldwide. Oxford is a registered trade mark of
Oxford University Press in the UK and in certain other countries

Published in the United States of America by Oxford University Press
198 Madison Avenue, New York, NY 10016, United States of America

British Library Cataloguing in Publication Data
Data available

Library of Congress Control Number: 2023931431

ISBN 978–0–19–886055–6

DOI: 10.1093/oso/9780198860556.001.0001

Printed and bound by
CPI Group (UK) Ltd, Croydon, CR0 4YY

Contents

Acknowledgments

I am extremely grateful to the many friends and colleagues with whom I've discussed the ideas in this book, some of whom read substantial portions of it in draft. I would like to thank Bob Beddor, Sam Bertsler, Ray Briggs, Jennifer Carr, David Chalmers, Tim Crane, Robin Dembroff, Kevin Dorst, Kenny Easwaran, Simon Goldstein, Ned Hall, Verity Harte, Sophie Horowitz, Harvey Lederman, Anna-Sara Malmgren, Alex Meehan, Laurie Paul, Agustín Rayo, Miriam Schoenfield, Declan Smithies, Jack Spencer, Jason Stanley, Scott Sturgeon, Zoltán Gendler Szabó, Roger White, Robbie Williams, Tim Williamson, Seth Yalcin, and especially Brian Hedden, who gave me extensive comments on an initial complete draft and *then* a rewritten one. Thanks also to two referees from Oxford University Press for their insightful and detailed comments. And thanks to Paul Forrester for insightful substantive comments, and for catching an embarrassingly large number of typographical errors.

I presented material from this book at a number of conferences and universities. Thanks to audiences at the 2019 Epistemology Workshop at the National University of Singapore, the 2019 Goethe Epistemology Meeting at the University of Frankfurt, the 2019 Evidence and Knowledge Workshop at the University of Glasgow, the 2020 Ranch Metaphysics Conference, the 2022 Central Division meeting of the APA, the 2021 Ideal and Non-Ideal Epistemology virtual seminar series, the 2022 Northeast Normativity Workshop, the University of Cologne, Rutgers University, and the University of Stanford.

Thanks to the members of PHIL 762: Idealization and Model-Building in Science and Philosophy, who gave me extremely helpful feedback on some early draft material.

Thanks especially to Kareem Khalifa, Dan Singer, and Bob Stalnaker, who gave me invaluable comments at a manuscript workshop hosted by Yale University (thanks to Yale too.)

I am grateful for permission to use material originally published elsewhere. Chapter 3 is adapted from "Fragmentation and Coarse-Grained Content," in *The Fragmented Mind*, published by Oxford University Press. I also use shorter sections of "Cognitive Mobile Homes" (published in *Mind* in 2017) and "Iteration Principles in Epistemology: Arguments For" (published in *Philosophy Compass* in 2015) in Chapters 4 and 7, respectively.

And thanks of course to my family: Ivana, Ben, and Sam.

Introduction

In his recent book *As If: Idealization and Ideals* (2017), Kwame Anthony Appiah makes a case for the significance of idealization in philosophy:

> often in philosophy it is useful to stand back and take a broad view of a topic, knowing that real progress requires work with a narrower focus as well. I offer this book in that spirit, hoping that it will prove useful in encouraging further explorations of idealization in aesthetics, ethics, and metaphysics, as well as in the philosophy of mind, of language, of religion, and of the social and natural sciences. . . .
>
> My aim, then, is not so much to announce any startling discoveries as to persuade you that idealization matters in all the major areas of the humanities and the sciences and in everyday life, and to commend it as a topic of reflection and research. (pp. ix–x)

The present book is in the spirit of Appiah's commendation. Like Appiah, I think idealization is ubiquitous in philosophy and everyday life. My focus, however, will be on idealization in epistemology. Of course, epistemologists have been building idealized models for over a century now—representing belief change using the mathematical formalisms of probability theory and modal logic is a well established practice. But while the *practice* of constructing idealized models in epistemology is old, metaepistemological reflection on the significance of that practice is not.[1] In this book I'll try to demonstrate that the fact that epistemologists build idealized models isn't merely a metaepistemological observation that can leave first-order epistemological debates untouched. Rather, once we view epistemology through the lens of idealization and model-building, the landscape looks quite different.

"It's just a model" is sometimes used as a shield to blunt the force of an objection, but it's a shield that can seem to diminish its bearer. In using it, one seems to concede that one's sights are less lofty than they might be. "I'll leave to other, more ambitious thinkers, the task of coming up with the

[1] Though see Titelbaum (2012), Yap (2014), and Williamson (2017).

Idealization in Epistemology: A Modest Modeling Approach. Daniel Greco, Oxford University Press. © Daniel Greco 2023.
DOI: 10.1093/oso/9780198860556.003.0001

final, comprehensive philosophical theory of knowledge (or justification, or evidence, or rationality, etc.); I'm just building models." This can sound like an abdication of the mete and proper role of philosophy.

To the contrary, I'll argue that constructing idealized models is likely the best we can do as epistemologists. There are good reasons to suspect that once we start using epistemological categories like belief, knowledge, and confidence, we've already entered the realm of idealization and model-building. So a non-idealized, comprehensive, and complete theory of these phenomena may be an oxymoron—akin to a non-idealized theory of frictionless planes. We can object to a model of knowledge by pointing to a better model, but in the absence of a better model, the fact that a framework for epistemological theorizing involves simplifications, approximations, and other inaccuracies—the fact of its status as an idealized model—is not in itself objectionable.

Once we accept that theorizing in epistemological terms is inescapably idealized, a number of intriguing possibilities open up. I'll defend a package of first-order epistemological views that might otherwise have looked indefensibly dismissive of our cognitive limitations—a package according to which we know a wide variety of facts with certainty, including what our evidence is, what we know and don't know, and what follows from our knowledge. This package turns out to be much more plausible once viewed through the lens of idealization; apparently devastating objections can be finessed, and apparently ad hoc features of the package come to look natural and elegant.

I'll also argue that we can see a novel route to something like contextualism in epistemology once we adopt the methodological approach I urge. Traditional versions of contextualism are motivated by linguistic analogies—contextualists argue that "knows" behaves like other context-sensitive pieces of language such as "tall" or "rich,"[2] while their opponents point to striking dissimilarities.[3] By contrast, I don't focus on linguistic analogies between "knows" and other context-sensitive pieces of language. Rather, I focus on analogies between the theoretical roles of epistemological categories, and the theoretical roles of categories in sciences where a modeling approach is often viewed as attractive. The economist who uses different, incompatible models of the macroeconomy in different theoretical contexts, or for different practical purposes, needn't embrace any distinctive semantic theses about "money," "growth," or "inflation." The biologist who uses a plurality of models of species, genes, or organisms, needn't embrace any distinctive semantic theses about "species," "gene," or "organism." Similarly but more controversially, I claim,

[2] E.g., Cohen (1999). [3] E.g., Stanley (2005).

the epistemologist who uses different, incompatible models of knowledge in different theoretical contexts, or for different practical purposes, needn't embrace any distinctive semantic thesis about the word "knows."

In the remainder of this introduction I'll provide a brief outline of each chapter. Readers who want to jump right in should feel free to skip to Chapter 1.

In Chapter 1 I'll introduce the concepts of idealization and modeling, discuss their relation, and provide some examples of model-building in philosophy.

I claim that not only does epistemology involve modeling, but epistemological modeling should be *modest*. In Chapter 2, I'll explain what I mean by modest modeling, and will introduce and motivate the concept via examples from economics. In a nutshell, a modest modeler isn't trying to point the way towards some grand unified theory of the domain she studies. She'll be content with the idea that in some domains, a collection of models, each partial and less than fully accurate, is the best we should hope for.

From Chapter 3 onwards I'll turn to applications, starting with the case of using sets of possible worlds to model the contents of belief and knowledge. While the virtues of this approach are well known, it also faces a family of familiar objections. A familiar response to these objections involves what's known as "fragmentation," the idea that agents shouldn't be modeled as having a single set of propositional attitudes, but instead multiple, inconsistent, "fragmented" sets. This thought has seemed to many like a desperate, ad hoc maneuver made to save possible worlds theories of content from devastating counterexamples. To the contrary, I'll argue that we can motivate fragmentationist ideas from first principles, independently of the need to avoid refutation. I'll argue that, given a popular view about the nature of the mind—interpretivism, of the sort defended by Davidson, Lewis, Stalnaker, and Dennett—*all* propositional attitude ascription involves a kind of implicit model construction; to have a propositional attitude is to be such that a certain kind of interpreter trying to predict and/or explain your behavior would model you as having that attitude. And if our interpreter is a modest modeler, then she will be content using different models in different contexts. When engaged in one theoretical project, she might model some agent as having some set of propositional attitudes, but when engaged in other projects, she will model that very same agent as having different, incompatible attitudes.

Having motivated fragmentationist views of the mind via interpretivism and modest modeling, I'll discuss the relationship between fragmentation and contextualism. In short, once we are fragmentationists, a broadly contextualist theory of belief attribution is extremely natural. I'll call the sort of contextualism emerges "model contextualism," and will discuss its relationship to extant forms of contextualism in the literature.

In Chapter 4 I'll turn to the topic of certainty. Much epistemological modeling represents agents as certain of a variety of facts—logical truths, their evidence, whatever they know. But—so a common narrative goes—ever since the failure of the Cartesian project we've known that almost nothing is certain. This might seem like a natural place to start talking about idealization; it's fruitful for various purposes to represent facts as certain when, strictly speaking, they're merely highly probable. This is not the approach I'll defend. While I'll agree that we're idealizing when we represent agents as being certain, I'll argue it's too quick to assume that there are strict truths to the effect that what we represent as certain is instead merely highly probable.

In Chapter 5 I'll apply some lessons from Chapter 4 to questions about the relationship between folk psychological categories like belief, and decision theoretic categories like credence. In particular, I'll argue that the idea that folk psychology has a distinct and ineliminable role to play in explaining certain aspects of our cognitive lives—one that cannot be coopted by decision theory—looks a good deal less plausible once these lessons are appreciated. This is because decision theory *can* make liberal use of certainties, and these certainties can do the theoretical work often claimed to be the distinctive purview of folk psychological belief.

In Chapters 6 and 7 I'll shift focus to issues involving higher-order evidence and knowledge. It's common practice in epistemology to distinguish between claims at different epistemic levels. Just as we can ask whether I know that P, we can ask whether I know that I know that P, or whether I know that I know that I know that P, and so on all the way up. Likewise with other epistemological concepts, such as probability, evidence, and rationality. We can ask whether P is probable, but also whether it is probable *that* P is probable, or whether it's rational to believe that P, but also whether it's rational to believe *that* it's rational to believe that P. Questions about the relationships between epistemic levels turn out to be relevant to a wide range of epistemological debates. Certain skeptical arguments rely on strong claims about the links between different epistemic levels, and many philosophers have thought that we must reject these level-bridging claims if we are to avoid skepticism. More recently, debates about how we ought to respond to evidence of our own fallibility—for example, debates about the epistemic significance of disagreement—have often hinged on whether some level-bridging claim is true.

In Chapter 6 I'll argue that many cases that have been seen as illustrating divergences between first-order and higher-order epistemic statuses—cases where someone knows, but doesn't know that she knows, or has evidence, but lacks evidence *that* she has evidence—are better understood in other terms. We can model such cases in a framework in which higher-order intrapersonal knowledge and evidence are trivial, so long as we are modest modelers who accept that no *single* model in that framework can capture everything we might hope for.

In Chapter 7 I'll defend the fruitfulness of models including common knowledge—knowledge that we know that we know . . . ad infinitum—against some recent attacks. Some of those attacks—from Williamson, Hawthorne, and Magidor—turn on rejecting the sorts of principles concerning relations between first-order and higher-order knowledge that I defend in Chapter 6. Another, due to Harvey Lederman, does not; Lederman grants the KK thesis—if you know that P, you know that you know—for the sake of argument, and contends that even with that major concession, common knowledge is impossible. Ultimately, I'll argue that Lederman's strategy looks much less persuasive once we adopt a modest modeling perspective.

Finally I'll conclude in Chapter 8 by considering how the perspective developed in the preceding chapters bears on debates about the relationship between "ideal" and "non-ideal" theory in epistemology.

1
Idealization and Modeling

Both "idealization" and "modeling" are used in a bewildering variety of ways, across many literatures. As a first pass, we can distinguish between broader and narrower construals of these terms. Some writers distinguish idealization from abstraction, simplification, and approximation, while others use it as more of an umbrella term for acceptable innacuracy. Similarly, some writers use "modeling" as a term for a distinctive type of inquiry that some but not all scientists engage in,[1] while others use it as an umbrella term for the construction of representations that fit their targets only in limited respects or degrees; an activity all people—and a fortiori, all scientists—engage in.

My own use of these terms will be rather capacious. Here I'm following Giere (1988), who did as much as anyone else to inspire the "modeling turn" in the philosophy of science. Giere offered what he called a "cognitive" theory of science. The central analogy of his book was between the mental models studied by cognitive scientists, and scientific theories more generally:

> I will argue that scientific theories should be regarded as similar to the more ordinary sorts of representations studied by the cognitive sciences. There are differences to be sure. Scientific theories are more often described using written words or mathematical symbols than are the mental models of the lay person. But fundamentally the two are the same sort of thing. (p. 6)

Giere goes on to use "idealization" in a way that suggests that all models are idealized.[2] I don't deny that for many purposes it's fruitful to narrow in on more discriminating senses of "idealization" and "modelling". A research lab advertising to hire a modeler would be rightly frustrated by an applicant who insisted that he was qualified merely by virtue of having a working visual system, which after all builds models of his local environment. But for my purposes, like Giere's, the broad senses of these concepts will suffice. When these broad, undiscriminating senses of "idealization" and "model" are used,

[1] For example, Williamson (2017).
[2] E.g., "I suggest calling the idealized systems discussed in mechanics texts . . . "models" " (p. 79).

Idealization in Epistemology: A Modest Modeling Approach. Daniel Greco, Oxford University Press.
DOI: 10.1093/oso/9780198860556.003.0002

the claim that some representation is an idealized model will tend to be unsurprising. But unsurprising claims can still have surprising consequences. Ultimately, I'll argue that once we appreciate that epistemology is shot through with idealization—even in this broad sense of "idealization"—the argumentative landscape ought to look quite different.

Michael Weisberg (2013, p. 4) characterizes modeling as "the indirect study of real-world systems via the construction and analysis of models." Of course, the natural question to ask upon hearing such a definition is, "what are models?" Here definitions are perhaps less helpful than examples. Models can be, as the name suggests, concrete physical objects. Weisberg describes an example in which the Army Corps of Engineers constructed a massive scale model of the San Francisco Bay in a warehouse, complete with hydraulic pumps to simulate tidal and river flows, in order to study the potential effects of damming the San Francisco Bay. But models can also be abstract mathematical objects. A government interested in selling rights to sections of the broadcast spectrum faces lots of questions about how to do this. It's typically done via auction, but auctions can be designed in a wide variety of ways. Should parties know each other's bids? Should bidding occur simultaneously or sequentially? Should bidding itself cost money, even for non-winning bidders? Should the highest bidder pay her bid, or perhaps the second highest bid? These questions and others could perhaps in principle be studied in a direct empirical fashion, by running lots of auctions under various different designs and analyzing the results. But this may be impossible, or practically infeasible. For instance, there may not be enough public assets the government is willing to auction off to make it possible to get the data necessary to directly study a wide range of auctions. It's much more feasible to hire economists to construct game theoretic models—abstract descriptions involving a set of agents, a set of states of the world, the values that the agents assign to those states, strategies available to the agents, and rules for predicting the agents' actions on the basis of the former—in order to answer questions about the likely effects of these design choices.[3] Whether the models we construct are concrete or abstract, the advantage of the approach is that it's easier to study a model than it is to (directly) study the real-world system it represents.[4]

[3] See Klemperer (2004) for an overview of the theory, and McAfee and McMillan (1996) for some discussion of how that theory has been used by the FCC.

[4] An astute reader might already notice an apparent contradiction. The characterization of modeling just offered suggests a contrast between modeling (where we study the world indirectly via models), and some distinct strategy where we study the world directly. But how is that consistent with the idea that we're discussing modeling in such a broad and capacious sense that just about all representation

The examples just given, along with the general definition of modeling, suggest that it is a purely descriptive activity—one models a real world system to find out how it actually is, or how it would change under various interventions. But I also want to allow for the possibility of normative modeling, where the aim isn't to generate descriptions, but evaluations, recommendations, or prescriptions.[5] If I'm facing some difficult decision, I might write down a pro and con list with the aim of generating a prescription for what *to* do, rather than a mere prediction about what I *will* do. Despite this difference, it's illuminating to think of the list as akin to the models discussed in the previous paragraph. There's some real world system—me and my practical situation—that I reason about indirectly by dealing with a more tractable representation of it.[6] That the reasoning concludes in a prescription rather than a description is immaterial to whether it should be thought of as modeling.

A central fact about modeling is that models are not intended to capture every aspect of the systems they model. The model bay constructed by the army corps of engineers didn't have any fish, nor was it as big as the actual San Francisco Bay. Game theoretic models don't specify every detail of the strategic situations they model; the heights, eye colors, and clothing of the players in the game aren't represented. And pro and con lists aim to pick out the most central or significant positive and negative considerations pertaining to a course of action. If I'm trying decide whether to adopt a dog, I won't list as a pro the savings in my heating bill that would result from an extra warm body in the house. That models leave out aspects of the systems they model is not an accident—if models were just as detailed and complex as the systems they're used to model, then they would be no more tractable than those systems, and

amounts to modeling? Here I'll appeal to the broadly Kantian, indirect realist thought that we never study the world directly—even the most basic perceptual experience is fruitfully viewed as involving the construction and analysis of (mental) models, as attested to by a rich tradition in cognitive science. See Clark (2016) for a survey. So perhaps a more precise but less pithy way of putting the point in the main text would be that it's often easier to study a system indirectly via the explicit and intentional construction of (concrete or abstract) models than it is to study it more directly (but *still* indirectly!) via the implicit construction and analysis of mental models.

[5] Titelbaum (2012) explicitly characterizes his project as a kind of normative modeling. See also Colyvan (2013).

[6] Of course, questions about the relationship between practical reasoning about what *to* do, and prediction about what I *will* do, are controversial. Some writers—e.g., Velleman (1989)—may think of them as more closely related than the discussion in this paragraph suggests. For such writers, normative modeling—at least in the first-personal case—may look hard to distinguish from descriptive modeling.

the point of modeling would be lost. A map that is as detailed as the physical territory it maps is far too cumbersome to use for navigation.[7]

1.1 Models and Frameworks

Following Titelbaum (2012), I find it helpful to distinguish between models and modeling frameworks. A model is a particular representation, the study of which is intended to illuminate some specific system or object. A modeling framework is a general recipe, approach, or set of techniques for building models of the same kind. For example, the London Tube map is a model of the London underground transit system. But we can also describe a more general modeling framework used to construct similar maps. It would involve the following rules, among others. When modeling a transit system, use different colors for each line. Make sure to represent each station, and to make it clear which lines can be accessed at which stations. Don't worry too much about geographical accuracy—legibility is more important than getting relative distances or the shapes of land masses just right. Include major landmarks—blue for rivers, green for large parks, but don't label them, and certainly don't label streets.

Ultimately, I want to suggest that when certain epistemological positions—the package of "Cartesian" views I'll describe in later chapters—are viewed as modeling frameworks, what might have otherwise looked like powerful objections lose much of their force. But to be able to do that, it will be helpful to start with some more general reflections on the virtues modeling frameworks can have, and what we can reasonably ask of them.

[7] Versions of this example have been used by a wide variety of thinkers across many disciplines, including Lewis Carrol and Umberto Eco. My favorite illustration is a short story by Jorge Luis Borges, "On Exactitude in Science," which appears in its entirety below:

> In that Empire, the Art of Cartography attained such Perfection that the map of a single Province occupied the entirety of a City, and the map of the Empire, the entirety of a Province. In time, those Unconscionable Maps no longer satisfied, and the Cartographers Guilds struck a Map of the Empire whose size was that of the Empire, and which coincided point for point with it. The following Generations, who were not so fond of the Study of Cartography as their Forebears had been, saw that that vast map was Useless, and not without some Pitilessness was it, that they delivered it up to the Inclemencies of Sun and Winters. In the Deserts of the West, still today, there are Tattered Ruins of that Map, inhabited by Animals and Beggars; in all the Land there is no other Relic of the Disciplines of Geography.
>
> (purportedly from Suárez Miranda, *Travels of Prudent Men*, Book 4, ch. XLV, Lérida, 1658)

A modeling framework isn't the sort of thing that can be true or false, so one can't defend a framework by arguing that it's true. The most straightforward way of defending a modeling framework is to use it to construct some models, and to then argue that those models have the sort of virtues we want models to have. In the case of the transit maps, it's easy to see how this might go. I might use the framework described above to construct maps of various transit systems, and then point out that the maps fulfill their purpose of helping travelers efficiently plan their trips; if my maps help travelers get where they want to go, then that is good evidence that omitting street names is a harmless idealization. Formal or abstract modeling frameworks can sometimes be defended in the same way. If the purpose of some game theoretic modeling framework is to aid governments in designing auctions that will maximize revenue from selling off assets, then the framework can be defended by showing that governments using the models it produces have raised more revenue than governments using other methods; if a game theoretic modeling framework is fruitfully employed in auction design, that is good evidence that the various details its models leave out are not essential to the phenomenon it models.

This sort of strategy is easiest when the models the framework is used to construct have practical or predictive purposes whose success is simple enough, at least in principle, to judge. Some examples of model-building in philosophy may fit this mold. When philosophers build models to study the effects of communication and reward structures in science,[8] or political polarization,[9] or the evolution of language,[10] while it may be difficult in practice to obtain the sort of evidence that would tell strongly in favor of the fruitfulness of one or another model, it seems clear enough what sort of evidence could, in principle, do the trick. So while it may be difficult to *determine* whether some idealization is harmless, we can at least have a reasonably precise idea of *what it would take* for an idealization to be harmless.

But often the models philosophers build don't have simple practical or predictive purposes for which success is, at least in principle, easy to judge. In such cases, how can we judge whether a model is a good one—in particular, how can we decide whether omitting some detail is an acceptable simplification, or instead involves abstracting away from the very phenomenon the model

[8] See, e.g., Kitcher (1990), Strevens (2006), Bright (2017), Romero (2017), Zollman (2018), and O'Connor and Weatherall (2019).

[9] See, e.g., Bramson et al. (2017) and Singer et al. (2019).

[10] See, e.g., Lewis (2002/1969) and Skyrms (2010).

is designed to illuminate?[11] I don't have a general answer to this question. Before saying anything specific about how we might answer it in the case of philosophy, I'll note that we do have partners in guilt in the form of economists. Some economic models are designed for very concrete, predictive, practical purposes, but certainly not all. Some are "caricatures," in the sense memorably described by Gibbard and Varian (1978); caricatured models are designed "not to approximate reality, but to exaggerate or isolate some feature of reality." Many of the models used in epistemology are relevantly similar. Consider the "reflection principle,"[12] which says, roughly, that changes in one's level of confidence in a proposition should not be predictable. The models used to vindicate this principle are highly idealized; they involve assumptions to the effect that you're certain you'll be rational in the future, that you won't become certain in anything false, and more.[13] But the principle, I think, is fruitful in much the same way that the caricatured models of economics can be. If you find yourself expecting to be more confident in some proposition tomorrow than you are today—after you've inspected some evidence or considered some argument in its favor, perhaps—it's very likely that reflection on the reflection principle will convince you that you're making a mistake. If you expect that there's persuasive evidence that you haven't yet examined in detail you should *already* adjust your confidence upward, and then be prepared to lower it back down if, when you get a closer look, the evidence is less compelling than you'd expected it to be.[14] Just as the economist whose intuitions have been trained by caricatured models will be quicker to spot perverse incentives or comparative advantages, similar benefits accrue to those raised on a healthy diet of epistemological models.

More generally, I think we can often recognize examples where model-building in a philosophical context is genuinely illuminating—where a modeling framework seems to reveal something important about some phenomenon of philosophical interest, and adds to our understanding of some domain—despite the models it builds being incomplete and idealized in a host of respects, as well as lacking straightforwardly testable consequences.[15] So rather

[11] See Mills (2005) for some arguments that much political philosophy involves idealizing away from the phenomena it ought to be concerned with, and Beaver and Stanley (2019) for a similar argument concerning philosophy of language.

[12] Originally proposed by van Fraassen (1984).

[13] See Briggs (2009) for a helpful discussion.

[14] See Salow (2017) for a more extensive discussion.

[15] The question of how idealized models can contribute to our understanding is a much debated one. Some writers argue that it requires revisionary, non-factive conceptions of explanation or understanding (Elgin, 2007, 2009). My own sympathies are more in line with those of Khalifa (2017) who defends the idea that we can stick with traditional, factive conceptions of explanation and understanding

than offer a general, principled answer to the question of how we can evaluate model-building that doesn't have a narrowly predictive purpose, I'll provide a few more examples in which I hope it's relatively uncontroversial that model-building in philosophy is illuminating, even if it's hard to say exactly what entitles us to that verdict.

1.1.1 Absolute and Incremental Confirmation

Much philosophy involves the presentation and attempted dissolution of paradoxes. Model-building can play a crucial role in this process. Carnap's solution to one of Hempel's famous paradoxes of confirmation provides a nice illustration of how this can work. A general theory of confirmation—the relationship between evidence and the scientific hypotheses it supports—was the holy grail of mid-twentieth-century philosophy of science. Much of this work involved proposing putative principles governing the logic of confirmation. While Hempel (1945) discussed many such principles, I'll focus on just two of them:

Special Consequence Condition: If E confirms H, and H′ is a consequence of H, then E confirms H′.

Converse Consequence Condition: If H entails E, then E confirms H.[16]

Each principle has a good deal of intuitive plausibility. The special consequence condition seems to capture something important about how science works—once we get solid evidence for a theory, we draw out consequences of the theory to make predictions, offer explanations, and so on. That is, the very evidence that justifies us in believing a theory also justifies us in believing what we go on to deduce from that theory. If evidence could support a theory without supporting its consequences, then it's hard to see how this could work. And the converse consequence condition also seems plausible. How do we test a theory? We figure out what we should expect to see if the theory is true, and then we check to see whether reality matches those expectations. If it does—if the theory said we should expect to see some piece of evidence, and

while still allowing a role for idealized models. Roughly, when an idealized model explains some phenomenon, it's the genuine fact that the model is *approximately* true, or accurate, that we can plug into traditional conceptions of explanation. But I think that I can remain largely neutral on this question for my purposes here.

[16] These are slightly adapted from Hempel (1945, p. 102).

that's exactly what we observe, then we've just gotten some confirmation that theory is true. The converse consequence condition looks to encapsulate this plausible idea.

And yet together they have the consequence that, for any two claims A and B, A confirms B. Let A be some arbitrary piece of observed evidence, e.g., snow is white. Let B be some unrelated claim for which A provides no evidence, e.g., dogs hate peanut butter. Now, introduce the conjunction A & B, that is, snow is white and dogs hate peanut butter. A & B entails A, so by the converse consequence condition, A confirms A & B. That is, snow is white confirms snow is white and dogs hate peanut butter. That conjunction entails that dogs hate peanut butter—A & B entails B. So by the special consequence condition, because A confirms A & B, and A & B entails B, A confirms B. But this is absurd—the fact that snow is white provides no reason to accept that dogs hate peanut butter.[17]

Hempel's reaction was simply to deny the converse consequence condition. But while that response avoids absurdity, it does so at a cost; as already noted, the converse consequence condition seems to capture something important about hypothesis testing. It would be more satisfying, then, to be able to see it as containing at least a grain of truth. This was Carnap's (1950) approach. Following Aquinas' famous advice—"when you meet a contradiction, draw a distinction"—Carnap distinguished two different senses of confirmation. According to Carnap, each of the two Hempelian principles is true of one, but not the other, sense of confirmation. Very roughly, evidence **incrementally** confirms a hypothesis when it provides *some* support for that hypothesis. Evidence **absolutely** confirms a hypothesis when it *establishes* the hypothesis—when the hypothesis is rationally acceptable in light of the evidence. While this verbal gloss is, I hope, somewhat helpful, illuminating this distinction is where Carnap's modeling framework shines. Once we model the relationships between evidence and hypotheses using the tools of probability theory, as Carnap did, the distinction becomes crystal clear.

In Carnap's modeling framework, incremental confirmation involves raising the probability of a hypothesis; E incrementally confirms H when $P(H \mid E) > P(H)$. And in general, the converse consequence condition holds true of it. If H entails E, then the conditional probability $P(H \mid E)$ will be greater than the unconditional probability $P(H)$.[18] But incremental confirmation, understood probabilistically, does *not* satisfy the special consequence condition. A piece

[17] In fact, dogs love peanut butter.
[18] Subject to the constraints that $P(H) \neq 0$, and $P(E) \neq 1$.

of evidence can raise the probability of some hypothesis without raising the probability of each of the consequences of the hypothesis. If I learn that your pet has no hair, that raises the probability that your pet is a sphynx cat. And of course, if your pet is a sphynx cat, that entails that it's a cat. But learning that your pet is hairless does *not* raise the probability that your pet is a cat.[19]

By contrast, E absolutely confirms H when the probability of H given E is above some threshold, 0.9, say.[20] The special consequence condition *does* hold of absolute confirmation. If $P(H \mid E) > n$, then $P(H' \mid E) > n$ for any H' entailed by H; if it's at least 90% likely that your pet is a sphynx cat, then it's at least 90% likely that your pet is a cat. But the converse consequence condition does *not* hold of absolute confirmation. It's easy to construct models where learning that your pet is hairless raises the probability that it's a sphynx cat, but still leaves that probability well below 90%.

Stepping back, Hempel left us with a pair of claims about confirmation, both of which looked attractive, but which weren't jointly acceptable. While Hempel simply rejected one of them, Carnap did better; he gave us a modeling framework for thinking about evidence and hypotheses that let us cleanly distinguish two different senses of confirmation, allowing us to see a grain of truth (and also of falsity) in each of Hempel's principles.

Moreover, I claim that this insight would survive even if Carnap's probabilistic framework for modeling support relations between evidence and hypotheses turned out to be incomplete or misleading in some or another respect.[21] It's possible to use a framework to illustrate a distinction—and thereby dissolve a paradox—even if the framework is imperfect. One way to see this is to note that a host of alternative frameworks to Carnap's can dissolve Hempel's paradox in much the same fashion. For example, a Dempster-Schafer theorist, while using a very different set of formal tools from Carnap, can avail herself of essentially the same incremental/absolute distinction.[22]

Why is it fruitful to view Carnap's contribution through the lens of idealization and modeling? Carnap (1950) himself described his methodology in terms of "explication," and used the word "model" only in its technical, logical sense. Nevertheless, it strikes me as fruitful to view Carnapian explication as a species of model-building. For Carnap, explication involved taking an inexact, prescientific concept—the explicandum—especially one beset by

[19] The example is lightly adapted from Pryor (2004).

[20] For our purposes, nothing turns on the choice of threshold.

[21] As most philosophers, myself included, would agree it was.

[22] See Halpern (2003), esp. ch. 2, for an overview of various frameworks for representing uncertainty, including Dempster-Schafer theory.

incoherence and paradox, and proposing an exact, rigorously defined and coherent replacement—the explicatum. He emphasized four criteria a good explicatum must meet: similarity to the explicandum, exactness, fruitfulness, and simplicity. Both the Carnapian explicator and the contemporary modeler start with some phenomenon (an explicandum, or a target system) that they'd like to better understand. They attempt to better understand that phenomenon by constructing a surrogate (an explicatum, or model), which has the virtues of being more accessible (exact and simple, or tractable) than the phenomenon they started with, while still similar enough in the right respects to the original phenomenon for analysis of the surrogate to amount to making progress on the original epistemic goal.[23]

Of course, there are important differences. Carnap's explicanda are always concepts, whereas the target systems of models *might* be conceptual, but in general are not; a meteorologist who builds models isn't trying to learn about concepts, she's trying to learn about the weather. Moreover, because Carnap tends to talk about explicata as *replacements for* explicanda, rather than *representations of* explicanda, he doesn't conceive of dissimilarities between explicata and explicanda as inaccuracies. But these differences strike me as overshadowed by the similarities. In practice, a Carnapian who thinks of the probability calculus as (part of) an explication of pre-theoretical notions like evidence and support, and a modeler who thinks of the probability calculus as (part of) a model of rational support relations (e.g., Titelbaum (2012)), may end up using their formalism in identical ways, to generate identical recommendations concerning the same problems. So I'm happy to count Carnapian explication as falling under the umbrella of idealization and modeling, in the broad and capacious senses in which those terms are used in this book.

While I've picked just one of Hempel's famous paradoxes to illustrate the illuminative power of probabilistic frameworks for modeling evidential support, the literature is filled with many more such examples. To gesture at just a few, the Ravens paradox,[24] Hume's problem of induction,[25] and the surprise examination paradox[26] have all been subject to attempted—and in my view, sometimes quite satisfying—dissolution by probabilistic modeling.

[23] See also Novaes and Reck (2017), who discusses the relationship between Carnapian explication on the one hand and the idea of formalisms as cognitive tools on the other.

[24] See Fitelson and Hawthorne (2010) for a survey of Bayesian treatments of the Ravens paradox, along with Rinard (2014) which in my view improves on prior treatments.

[25] See Okasha (2001, 2005).

[26] See Hall (1999) and Kim and Vasudevan (2017).

1.1.2 Propositional Logic

Even in the absence of some paradox to be dissolved, modeling frameworks can play a role in illuminating similarity or shared structure. Take the formal language of propositional logic. This can be studied in its own right as a mathematical object—it can be the subject of soundness and completeness proofs, for example. But it can also be used as a tool for studying natural language. A standard exercise in introductory logic classes involves taking a piece of natural language argumentative prose—perhaps a newspaper op-ed or an excerpt from a philosophical text—"translating" it into the language of propositional logic, and then evaluating the translated argument for soundness and/or validity. This exercise can be fruitfully thought of as a kind of normative model-building.[27] We're interested in evaluating an argument couched in natural language. But we take an indirect route. We build a model of the natural language argument in the language of propositional logic, and then engage in the comparatively straightforward task of evaluating the model.[28] Of course, this approach involves abstracting away from a lot—sometimes so much that it's misleading; there are good arguments whose goodness isn't revealed by this method.[29] Nevertheless, the framework can be genuinely illuminating. When a student sees two superficially dissimilar pieces of natural language argument as having the same formal virtue—both are instances of modus tollens, perhaps—she has achieved a philosophically significant insight, despite having used idealized, incomplete models to do so.

Just how widespread is model-building in philosophy? While I've given a few clear examples from relatively technical parts of the discipline, in my view there is also a great deal of informal modeling. In fact, I'm sympathetic to a general view about reduction and emergence, inspired by Dennett (1991), explicitly generalized by Wallace (2012), and also arguably endorsed by Carroll (2016), to the effect that *all* discourse about non-fundamental aspects of reality—basically, everything but quantum mechanics—is fruitfully understood in terms of modeling. Just as classical mechanics is useful—when it is—because certain quantum systems can be tractably modeled as classical systems without too much loss of accuracy, so too with biology, psychology, and the everyday ontology of medium sized dry goods. I don't pretend that this is an adequate summary; I'll expand on this idea in the next chapter.

[27] See Titelbaum (2012).

[28] Straightforward at least if all we're interested in is validity. Soundness, of course, is harder.

[29] For instance, arguments whose validity is only revealed by moving to more sophisticated logics, or arguments that are not formally valid in any logic, but are still substantively reasonable.

1.2 Rejecting Models

Just as one can't defend a modeling framework by arguing that it's true, one can't attack a modeling framework by arguing that it's false. But that doesn't mean that "it's just a model" can be legitimately used as an all-purpose shield against objections; there are bad modeling frameworks, and bad models. Perhaps most obviously, modeling frameworks can be useful for some purposes, or in some situations, but not others. In mechanics, models that leave out air resistance and frictional forces provide nice, simple explanations of why, *pace* Aristotle, bowling balls and basketballs will fall at roughly equal speeds, despite the former weighing much more. But using such models to predict the trajectory of a falling feather will only lead to disappointment. In philosophy, propositional logic occupies a similar role. Some natural language arguments are fruitfully modeled in propositional logic. But not all; arguments whose validity depends on their quantificational structure or on the use of modal vocabulary are not best modeled in that framework. In both of these cases, negatively evaluating the models these frameworks build is relatively straightforward. A framework that ignores air resistance, when used to model the descent of a feather, will make false empirical predictions. And propositional logic, when used to model arguments with quantificational structure, will render false verdicts about validity. What these examples illustrate is that a modeling framework can be useful and illuminating despite having less than universal scope, even in what might have looked like its natural domain. Not all natural language arguments are fruitfully modeled in propositional logic. Not all dynamics problems are fruitfully modeled in a framework that ignores air resistance. But some of them are, and that's enough for the frameworks to be worth keeping around.

This point can be a bit surprising, so it's worth dwelling on. Given that there are modeling frameworks that take account of air resistance, and so make correct predictions concerning both bowling balls *and* feathers, why is it ever illuminating to work with models that ignore it? And given that there are more sophisticated logics that can represent quantificational, predicate, and modal structure, why should we ever model arguments using simple propositional logic? This is a big question, but I think the broad outline of the answer is relatively uncontroversial. A good explanation doesn't include unnecessary detail. In an ideal explanation, the explanans is as detailed as it needs to be to account for the explanandum, and no more.[30] This isn't

[30] See especially Yablo (1992).

merely a matter of convenience. If I explain why somebody died by saying that they were hit by a lime green, all-wheel-drive 2013 Subaru Forester, I haven't merely been unnecessarily verbose. Because the death didn't depend on the color, suspension, model, manufacturer, or year of the car, including all these details makes my explanation worse—it amounts to citing factors in my explanation that made no difference to the phenomenon I'm trying to explain. For similar reasons, in explaining physical phenomena that don't depend on friction or air resistance, models that leave out those factors aren't merely more convenient to work with that models that include them; rather, they provide better explanations that more faithfully reflect the structure of the phenomena they're meant to model. Likewise when it comes to using propositional logic rather than more sophisticated formal tools to explain what various instances of modus tollens have in common.

Of course, some frameworks *aren't* worth keeping around—they're not the right tools for *any* job.[31] The history of science and philosophy furnishes us with plenty of examples. Ptolemaic astronomers had detailed models of the solar system that did a more than serviceable job of predicting observed planetary motions. But today, heliocentric models are both simpler and more accurate than their geocentric predecessors. While the heliocentric framework is of genuine antiquarian interest, there are no phenomena best explained by appeal to the models it builds. Examples from philosophy are bound to be more controversial, but arguably the relationship between the framework suggested by Aristotle's writings about predication on the one hand, and the Fregean predicate calculus on the other, is similar. Plausibly, any explanation of why some natural language argument is valid that could be given by an Aristotelian logician could be given more perspicuously by a Fregean, and the Fregean would have the advantage of being able to model a wider range of logical phenomena.

The moral here is a broadly Kuhnian one, with modeling frameworks playing the role of paradigms.[32] According to Kuhn, mere anomalies, or recalcitrant data, aren't sufficient justification to reject a scientific paradigm. Instead, a better paradigm is needed. By analogy, to reasonably reject a modeling framework tout court—to dismiss it as incapable of providing *any* helpful or illuminating explanations, rather than just to regard it as inapplicable to certain situations—one must have in hand a strictly better framework. This is a theme

[31] I borrow the metaphor of models as tools, and the modeler as akin to a handyman with a diverse toolkit, and who uses different tools for different jobs, from Veit (2020), and I'll return to it throughout the book.

[32] See Kuhn (1962).

I'll return to throughout the book. I'll argue that various modeling frameworks in epistemology—ones that incorporate "logical omniscience" assumptions, or which model learning by the acquisition of certainties, or which "conflate" first-order and higher-order epistemic statuses, have virtues not (yet) matched by any competitor frameworks. And so despite "anomalies"—cases in which those frameworks seem to deliver awkward results—they shouldn't be put out to pasture.

I expect that not much I've said about modeling thus far will be particularly controversial. In part, this is because my remarks have been consistent with the view that modeling, in both science and philosophy, has a kind of second-class status as compared with theory-building. On this picture, the proper aim in any domain of inquiry involves coming up with a fully general and accurate theory—a set of laws or principles that (perhaps together with some particular matters of fact, such as initial conditions) captures the whole truth about the domain. Model-building, in which we expect and accept a degree of mismatch between our models and the reality they represent, may be of use in various ways—perhaps for making predictions, or as a pedagogical aid, or as a preliminary to theory-building—but it's not the ultimate goal; proper inquiry is more ambitious than that.

In the next chapter I'll describe a view that rejects this ambition. On this alternative view, in many domains developing more and better models and the judgment to know which ones to use when is the best we can hope for. This move is not novel—philosophers of science have been defending similar positions for decades, and the idea that much or even most of science is better viewed as model-building rather than as a search for grand unified theories is by now probably entrenched enough to count as the received view.[33] But in my opinion, epistemologists haven't quite caught up.

[33] See, e.g., Cartwright (1983), Teller (2001), Batterman (2009), Dupré (1993), Mitchell (2003), Wylie (1999).

2
Modest Modeling

In this chapter I'll distinguish two ways we can see model-building as fitting into inquiry. The first approach, which I'll call "ambitious," involves aspiring—at least in principle—to the attainment of a complete and fully accurate theory of the domain being modeled, and hoping that the models one builds in the meantime will provide some insight into what that merely hypothetical theory might look like. To put it metaphorically, the ambitious modeler thinks that a perfect model exists—God's model—and hopes that the imperfect, idealized models she herself builds will provide some clues as to what God's model might look like.

On the second, "modest" approach to modeling, we can be agnostic or skeptical concerning the existence of a perfect model, and we don't need to think of the tractable models we construct and analyze as providing clues about its nature. Rather, the modest modeler is content to work with a collection of models, each partial and less than fully accurate, without holding out hope for a grand unification on the horizon. Modest modeling is closely related to what Weisberg (2013) has called "Multiple Models Idealization" (MMI) which he describes as "the practice of building multiple related but incompatible models, each of which makes distinct claims about the nature and causal structure giving rise to a phenomenon" (p. 103). A modest modeler thinks that MMI is the best we can do, while an ambitious modeler regards MMI as a sign of an immature discipline; progress will involve subsuming multiple models under fewer models, with a single model at the limit of inquiry.

Walter Veit (2020) has drawn a distinction between "model monism" and "model pluralism" that is very close to the distinction I intend between ambitious and modest modeling. For Veit, "model monists seek one model: the best model, the perfect model, the model that is general, precise, and realistic," while pluralists, by contrast, "see model diversity as an unavoidable feature of science," and embrace it (p. 95). Like me, Veit draws inspiration from Dani Rodrik's (2015) account of the methodology of economics, and he generalizes Rodrik's methodological lessons well beyond Rodrik's own discipline of economics.

Idealization in Epistemology: A Modest Modeling Approach. Daniel Greco, Oxford University Press. © Daniel Greco 2023.
DOI: 10.1093/oso/9780198860556.003.0003

As I intend these terms, one can support modest modeling in some domains, and ambitious modeling in others. Indeed, my read of the current state of the literature in philosophy of science is that modest modeling is typically treated as a reasonable default throughout the special sciences, while ambition is often thought of as appropriate in fundamental physics.

This combination of attitudes strikes me as exactly right. To put the rest of this book in a very small nutshell, I think the sort of modeling that goes on in epistemology and closely related subfields—modeling people as believers, knowers, desirers, possessors of probability and utility functions—is more like what goes on in the special sciences than it is like fundamental physics, and that modeling in epistemology should therefore be modest.

While this may sound obvious or uncontroversial—who would have thought epistemology was like physics?—it's often tacitly presupposed that there *are* canonical models in epistemology, with the same kind of generality and universality in the epistemic domain that fundamental physicists aspire to in theirs. For example, it's common (among epistemologists, at least) to think that not only is it illuminating to model this or that episode of learning by Bayesian conditionalization, but also that there is a single, canonical Bayesian model into which all possible episodes of learning could be fit.

To see what this might look like, consider the following example. Suppose you are playing Texas hold 'em poker, and you are holding two spades, while the "flop" has revealed another two spades. You're wondering whether one of the two cards yet to be revealed will be another spade; how likely are you to complete your flush? If we wanted to represent this situation with a Bayesian model, we'd normally use an event space corresponding to the various different cards that might be revealed as the game progresses, and where each new card revealed updates the probabilities by eliminating some of these possibilities. Doing so would reveal a ~35% probability that you'll complete your flush. If we are modest modelers, we may be content to leave things at that. By contrast, while the ambitious modeler may find this sort of model a perfectly serviceable practical guide—if you want to get good at poker, these are the models you should get comfortable with—she'll also think that in principle it could be embedded into a much grander one.[1]

This grand model would have as its outcome space not just the possible arrangements of cards in a deck, but all metaphysically (logically?) possible events. And the probabilities of these events wouldn't just be treated as

[1] I am alluding to the distinction between small world and grand world decision problems drawn by Savage (1954).

equal—as in the case of the equal probabilities of each possible state of a shuffled deck—but instead would be determined by their objective, intrinsic, prior probabilities.[2] In this grand model, the probability you'd complete your flush would be the result not of updating an outcome space involving cards on evidence about which cards have been revealed, but instead the result of taking the canonical, intrinsic prior probability function defined over all possible outcomes, and updating it on the sum total of the evidence propositions you've acquired and retained throughout your life, including the evidence you have about the game of poker you're currently playing. After that, the posterior probabilities for certain propositions in the grand model—that you're about to complete a flush—would closely match those of corresponding propositions in our little, local model.[3]

While the existence of such a canonical model is sometimes debated—sometimes just taken for granted—among epistemologists, this is usually treated as an in-house debate, to be settled by epistemologists. By contrast, I'm inclined to think the question looks very different when we zoom out a bit and think about epistemology as one among many disciplines that builds models. When we do so, I suspect that the view that there are canonical, universal models in epistemology—whether of learning, or of knowledge more generally—will come to look a bit quixotic. The kind of canonical status such an epistemologist hopes for is one which is typically regarded as a reasonable aspiration in fundamental physics, but nowhere else.

My strategy in this chapter will be as follows. First, I'll say a bit about what the "ambitious" approach to modeling in fundamental physics amounts to. I'll then give some examples of what a modest approach to modeling, in disciplines outside of physics, can look like. Lastly, I'll describe what I'll call the Dennett/Wallace/Carroll view about reduction and emergence, which I'll argue supports a default skepticism about the existence of canonical, universal models outside of fundamental physics.

[2] For examples of such views, see Swinburne (2001), Chalmers (2012) (see especially the "frontloading" argument in ch. 4), and Hedden (2015), among many others. The idea that there is such a model has been attacked. See, e.g., Titelbaum (2010). Though there and elsewhere—e.g., in the literature on "epistemic permissivism"—it's mainly just the existence of a unique, objective prior probability function that is disputed. The existence of a canonical outcome space is typically taken for granted by all parties to these debates. That strikes me as a mistake.

[3] I say "closely" on purpose. An ambitious modeler will likely think the probability that you'll complete your flush according to the little model is, strictly speaking, a bit too high. After all, God's model will allow for some small chance that the game will be cut short before all the cards are dealt—maybe a meteor will strike—in which case you won't complete your flush regardless of the state of the deck.

2.1 Ambitious Modeling

The late theoretical physicist and Nobel laureate Stephen Weinberg gives a nice statement of what I'm calling an "ambitious" approach to modeling:

> Our present theories are of only limited validity, still tentative and incomplete. But behind them now and then we catch glimpses of a final theory, one that would be of unlimited validity and entirely satisfying in its completeness and consistency. We search for universal truths about nature, and, when we find them, we attempt to explain them by showing how they can be deduced from deeper truths. Think of the space of scientific principles as being filled with arrows, pointing toward each principle and away from the others by which it is explained. These arrows of explanation have already revealed a remarkable pattern: they do not form separate disconnected clumps, representing independent sciences, and they do not wander aimlessly—rather they are all connected, and if followed backward they all seem to flow from a common starting point. This starting point, to which all explanations may be traced, is what I mean by a final theory. (Weinberg, 1994, p. 6)

Weinberg acknowledges that, even if his dreams of a final theory are realized, in practice we'll still need sciences other than fundamental physics—"different levels of experiences call for description and analysis in different terms" (1994, p. 62). But the overall picture he paints is one in which this reflects merely practical limitations on our part. God could give a perfectly accurate and complete description of the world using the categories of fundamental physics; finite creatures like us have to get by using less accurate, more parochial, but more tractable models drawn from disciplines like economics and biology.

Much recent work in metaphysics presupposes the existence, at least in principle, of something like a final theory. David Lewis tells us that the maximally natural properties are those properties that would figure in a completed physics, and it's clear that Lewis takes the completion of physics to involve the discovery of a Weinbergian final theory (1983a, pp. 364–5). And while Lewis' discussion of natural properties has been scrutinized and generalized, the link to physics has, as far as I can tell, remained relatively uncontested.[4]

Moreover, the idea of a "final theory" is not new. It goes as least as far back as Laplace. Recall his famous statement of determinism:

[4] See, e.g., Sider (2011) and Chalmers (2012).

> We may regard the present state of the universe as the effect of its past and the cause of its future. An intellect which at a certain moment would know all forces that set nature in motion, and all positions of all items of which nature is composed, if this intellect were also vast enough to submit these data to analysis, it would embrace in a single formula the movements of the greatest bodies of the universe and those of the tiniest atom; for such an intellect nothing would be uncertain and the future just like the past would be present before its eyes.[5]

In the course of making a claim *about* the canonical, universal, final model of physics—namely, that it would be deterministic—he's presupposing that such a model exists. That is, he's presupposing that there is a set of physical categories that suffice to give a complete, canonical description of physical reality.

While the idea that there exists a final theory in physics is not uncontested—as I interpret it, Nancy Cartwright's work represents a radical rejection of ambitious modeling throughout the sciences, including in physics[6]—it seems to me to have the status of the conventional wisdom, and I'm inclined to agree with it.[7]

2.2 Some Examples of Modest Modeling

Believing in the existence of a final, complete, canonical model in physics doesn't yet commit us to anything concerning whether such models exist in other domains. Ultimately, I hope you'll agree that our default should be skepticism concerning the existence of canonical models in areas very far from physics. I want to get to the point where the ambitious Bayesian from a few pages back looks like a starry-eyed utopian dreamer. But it will take awhile. Let's warm up with an example involving coastlines, before attempting to generalize.

2.2.1 The Coastline Paradox

In 1951, the mathematician and pacifist Lewis Fry Richardson set out to investigate what looks like an eminently tractable empirical question: does

[5] See Laplace (1951).

[6] See, e.g., Cartwright (1983) and Cartwright (1999).

[7] If you, reader, are inclined to reject this conventional wisdom in favor of Cartwright's position, then the arguments of this book should be easier, not harder, to accept.

the length of a border between two countries affect the probability that they'll go to war? But before even getting to analysis, he found there were considerable difficulties in describing the data. According to Spain's reports, the Spain-Portugal border was 987 km. According to Portugal's, it was 1214 km. According to the Netherlands' reports, the Netherlands-Belgium border was 380 km. According to Belgium's, it was 449.5 km. As he wryly summed up, "there is evidently no official convention which can hamper the intellectual discussion of lengths of frontiers."[8]

What was going on? Imagine you're measuring the border of a country with a ruler. You'll place the ruler along the border of the country, with the specification that both ends of the ruler must be on the border (you can't require that the entire ruler touch the border, because the border might not be perfectly straight). You'll go around the entire country, adding up the ruler-lengths to get your final answer. What Richardson found is that the measured length of a border is highly dependent on the length of the ruler. In particular, for sufficiently "wiggly" borders—given a mathematical measure of wiggliness he defined—as we use a shorter ruler, the measured length increases without bound.

This is often called the coastline paradox, because Mandelbrot (1967) expanded on Fry's discussion using the coastline of Britain as an example. Moreover, the phenomenon is especially dramatic in the case of coastlines.

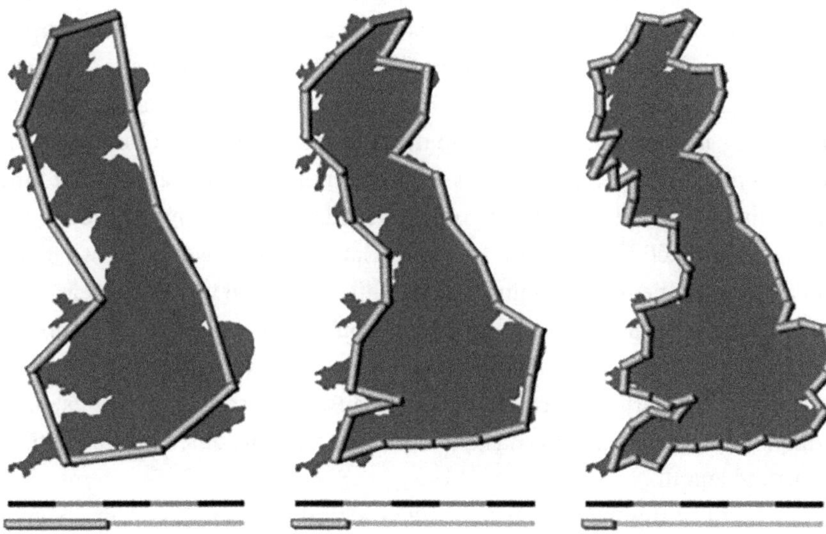

Figure 2.1 "An Illustration of the Coastline Paradox." From Wikipedia: https://en.wikipedia.org/wiki/Coastline_paradox

[8] Richardson (1993, p. 169).

What does any of this have to do with idealization and canonical models? We can regard any measure of a country's coastline as a kind of idealization. In adding up straight ruler lengths to come up with a number for the coastline, we're ignoring some matters of detail—the twists and turns smaller than the ruler—in order to get a tractable measure we can work with, e.g., for the purpose of formulating and evaluating hypotheses about causes of war. For some, not-very-wiggly curves (e.g., a perfect circle), we can regard this method of measuring length as approximating a single, canonical, true length. If you use smaller and smaller rulers to measure the perimeter of a perfect circle, your measurements will converge on its length. But real-world borders and coastlines are quite different; we can't regard measurements using smaller and smaller rulers as better and better approximating some true length.

This is for at least two reasons. First, if a coastline is wiggly, but that wiggliness only shows up at a high resolution, then measurements using a longer ruler will *not* approximate measurements using a shorter ruler. If we're ranking countries by coastline length, and the countries differ with respect to the scales at which they are most wiggly, then not only will the lengths increase as the ruler length decreases, but also the ordering will change. So even if we assume there is a true coastline length, we can't regard the measurements we actually take as giving us much by way of clues as to what it is.

But second, as we use a smaller and smaller ruler, the question of just what the coastline *is* becomes murky; as Mandelbrot (1967) put it, "as ever finer features are taken account of... there is no clearcut gap between the realm of geography and details with which geography need not concern itself" (p. 636). Imagine attaching a gps device to an ant, and somehow compelling the ant to walk along a country's coast; suppose the ant is directed to walk as close to the water as possible without getting its feet wet. In that case, it matters what time of day the ant is walking, because of tides. And should the ant walk along the sides of coastal cliffs, or stay on top? If the latter, how do we decide what's on top and what's the side (imagine sloping cliffs)? When it comes to measuring other quantities, these sorts of decisions won't make a big difference on the margin—whether my hair or nails count as part of me won't significantly affect my weight. But in the case of coastlines, because these sorts of decisions make a difference to how wiggly a coastline is, they make a big difference to the final measured length.

The point of this example is to serve as a kind of proof of concept. Sometimes, when we use an idealized model—in this case, when we ignore low-level twists and turns in representing a coastline as having some particular length—there isn't, even in principle, some non-idealized model using the same categories which we can regard our model as an approximation to. To talk about coastlines is already to have engaged in something very much like idealization and modeling; to inquire into the non-idealized, true length of the British coast is to commit a kind of category mistake.

But while the case of coastlines is fun, one can reasonably doubt whether it has any lessons for epistemological modeling; after all, what holds for coastlines does *not* hold for heights or weights. So in the next section I'll discuss what modest modeling—modeling without any ambitions of illuminating the way towards a grand, unified, canonical model—can look like in economics, which might seem of more direct relevance to epistemology.

2.2.2 Modest Modeling in Economics

Economists build models to study a wide range of phenomena. I've already mentioned auctions, but such varied topics as traffic flows, taxation, and cheating in Sumo wrestling have been modeled by economists.[9] On their face, these models are quite diverse—the best available models of auctions will not suggest any predictions concerning the effects on GDP of cutting interest rates, or the effect on employment of a minimum wage hike, and vice versa. Dani Rodrik, the Ford Foundation Professor of International Political Economy at the Kennedy School at Harvard University, takes this plurality at face value, defending a kind of modest modeling take on economics as a discipline:[10]

> Rather than a single, specific model, economics encompasses a collection of models. The discipline advances by expanding its library of models and by improving the mapping between these models and the real world. The diversity of models in economics is the necessary counterpart to the flexibility of the social world. Different social settings require different models. Economists are unlikely ever to uncover universal, general-purpose models. (2015, p. 11)

Moreover, according to Rodrik, we shouldn't expect a highly systematic story about which economic models are useful when. He writes that in economics, "good judgment is indispensable in selecting from the available menu

[9] On traffic, see Vickrey (1963) for a classic example. On taxation, see any macroeconomics text. And on cheating in Sumo wrestling, see Levitt and Dubner (2005).

[10] Cartwright herself defends a similar approach to economics (1999, ch. 6–7) and Rodrik repeatedly cites her as an inspiration for his views about methodology.

of models," and that while "evidence can provide some useful guidance for sifting across models . . . the process remains more craft than science."

A contrasting approach to the field involves taking "methodological individualism" as a guiding principle.[11] In the form in which it has been most influential in economics, this idea holds that group-level phenomena should be explicable, at least in principle, as the outcome of a collection of individual actions—usually, rational individual responses to incentives. Macroeconomic models should have "microfoundations." S. Abu Turab Rizvi, economist and provost of Lafayette College, in a paper ultimately pessimistic about the project, describes it as follows:

> The basic logic in strict microfoundations and in [General Equilibrium Theory], generally speaking, is to try to derive macroeconomic properties from assumptions on economic agents considered individually; that is, to see if macroeconomic regularities follow from microeconomic assumptions.

The demand for microfoundations is a kind of local reductionism, and sits nicely with what's sometimes called a "layer-cake" picture of the world. This picture incorporates ambition about modeling in fundamental physics, along with additional commitments. Not only must everything reduce, in some sense, to particle physics, but the reduction must go in stages. The social must reduce to the mental which must reduce to the biological which must reduce to the molecular which must reduce to the atomic which must reduce to the subatomic.[12] Or something along those lines. For my purposes, it's crucial to recognize that this layer-cake picture is not entailed by the existence of a final physical theory. One can coherently doubt whether all macroeconomic phenomena have microeconomic foundations, without doubting whether they have micro*physical* foundations. Rodrik provides a humorous anecdote that, in my view, helps illustrate why one might incline towards such a position:

> Economists' attachment to particular modeling conventions—rational, forward-looking individuals, well-functioning markets, and so on—often leads them to overlook obvious conflicts with the world around them. Yale University game theorist Barry Nalebuff is more world-savvy than most, yet even he has gotten into trouble. Nalebuff and another game theorist found

[11] The term is due to Weber, though it has a long history in debates about social scientific methodology. See Heath (2015) for an overview.

[12] See Oppenheim and Putnam (1958) for a relatively explicit statement of this view.

> themselves in a cab late one night in Israel. The driver did not turn the meter on but promised them he would charge a lower price at the end of the ride than what the meter would have indicated. Nalebuff and his colleague had no reason to trust the driver. But they were game theorists and reasoned as follows: Once they had reached their destination, the driver would have very little bargaining power. He would have to accept pretty much what his passengers were willing to pay. So they decided that the driver's offer was a good deal, and they went along. Once arriving at their destination, the driver requested 2,500 shekels. Nalebuff refused and offered 2,200 shekels instead. While Nalebuff was attempting to negotiate, the outraged driver locked the car, imprisoning his passengers inside, and drove at breakneck speed back to where he had picked them up. He kicked them to the curb, yelling, "See how far your 2,200 shekels will get you now."

The moral of the story is that the sorts of microeconomic assumptions that work well when modeling auctions, or commodities trading—and which figure so crucially in extant attempts to provide microfoundations for macroeconomic phenomena, as in the rational expectations framework[13]—don't work so well when modeling other sorts of interactions. So the prospects for a universal economic modeling framework look dim.

An ambitious economic modeler might hold that this is merely a consequence of the limitations of our present frameworks for modeling individual choice. With more accurate psychological theories, we'll be able to explain and predict why and when people sometimes act like *homo economicus*, and sometimes don't. Moreover, whatever success is enjoyed by macroeconomic models of aggregate phenomena will be explicable, in principle, as the result of individual interactions as described by some more precise and accurate successor to contemporary psychology. Perhaps. But at present, this amounts to little more than a fantasy. While some researchers are hopeful, the project of deriving macroeconomic predictions from non-classical, psychologically realistic microfoundations is barely even in its infancy.[14]

Of course, it's well beyond the scope of this book to attempt to settle the question of whether we should be optimsitic or pessimistic about the microfoundations project in economics. My aim in this section is just to make salient that this is an open question, not settled merely by being ambitious about modeling in fundamental physics.

[13] See Muth (1961) for the original statement of that approach.

[14] Though see Baddeley (2014) for some optimism on this score.

In the remainder of this chapter, I'll introduce the "Dennett/Wallace/Carroll" view of emergence as modeling, and will offer some general reasons for thinking that such a view supports thinking that modeling in domains far from fundamental physics should be modest by default.

2.3 Emergence as Modeling

Suppose we believe that when God wrote the book of the world,[15] she wrote it in the language of fundamental physics. She didn't say "let there be light," but rather "$i\hbar\frac{\partial}{\partial t}\Psi(\mathbf{r},t) = \hat{H}\Psi(\mathbf{r},t)$," or something along those lines.[16]

But *we* don't speak, write, or think in that language, so how can we manage to represent the world accurately? What's the relationship between the fundamental structure of the world on the one hand, and the categories and distinctions we use on the other? This is the question of *emergence*; how does the world we know and love emerge from the world of fundamental physics?

This is an immensely difficult question in the philosophy of science, and I couldn't possibly hope to settle it in this book.[17] Rather, my aim is to sketch a view that already exists in the literature—one to which I am sympathetic—and to draw out some of its consequences in epistemology. In particular, I'll argue that the view I'll call the Dennett/Wallace/Carroll view has as a consequence the result that we shouldn't expect there to exist models using the sorts of categories we use in epistemology that have the same kind of universality and canonicity of the as-yet-undiscovered final model of fundamental physics.

But in order to better explain that view, I'll start by introducing a competitor view about emergence—one that would *not* vindicate the idea of modest modeling as a default outside fundamental physics.

On the alternative view, emergence can be understood as a kind of logical consequence. High-level, less fundamental categories can be *defined* in terms of lower-level, more fundamental categories, such that from a sufficiently rich low-level description of a system, together with the right definitions of the high-level categories in terms of the lower-level ones, the correct high-level description of the system will follow.

There are no uncontroversial cases of this sort of emergence. Perhaps the best contender is the relationship between statistical mechanics and

[15] Sider (2011).
[16] This is Schrodinger's equation, the fundamental equation of quantum mechanics.
[17] See Sklar (1993), esp. ch. 9, for a useful survey.

thermodynamics—it's often claimed that, e.g., the temperature of a gas is the mean molecular kinetic energy of the molecules that make it up. In fact, the project of defining thermodynamic categories like temperature in more fundamental terms is philosophically fraught—see Sklar (1993), and Callender (1999)—and while it's widely accepted that statistical mechanics *somehow* explains the success of thermodynamics, the questions of just how this explanation works, and whether it involves anything like derivation via bridge-definitions, are much more controversial.

But *if* this sort of view were right—if, in general, high-level descriptions of a system were true insofar as they logically followed from low-level descriptions together with the right definitions—then it's hard to see how there could fail to be canonical high-level descriptions of the world, given the existence of a canonical fundamental description of the world. Suppose that there is a unique complete description of the world in the language of fundamental physics—the wave function of the universe, let's say. On the picture of emergence currently being contemplated, that description of the world, when supplemented with the right definition of, say, particles in terms of the wave function, will give rise to a unique model of particle physics. And when that model of particle physics is supplemented with the right definition of molecules in terms of particles, we'll get our unique model of chemistry. On this picture, not only does God have a model of fundamental physics, but for any non-fundamental set of categories, there will be a unique, maximally accurate way of modeling the world in terms of those categories—one that will follow with the inevitability of logic from the fundamental description of the world, together with a long series of bridging definitions.[18]

What's the (or, rather, an) alternative? Rather than thinking of less fundamental descriptions as logical consequences of more fundamental ones, we can think of them as *good models* of more fundamental ones. This is the view I find, in one form or another, in Dennett (1991), Wallace (2012), and Carroll (2016), and to which I'll now turn.

In the previous chapter we saw cases in which some real world system—e.g., the San Francisco Bay, or a collection of individuals bidding on some good—can be tractably studied by the construction and analysis of models. If you want to know what would happen if we were to dam the San Francisco bay, a good strategy is to dam the physical model and see what happens. If you want

[18] This picture is suggested by Oppenheim and Putnam (1958). Moreover, while he doesn't endorse anything quite like the above, David Lewis (1983b) does talk of "derived laws that follow fairly straightforwardly" from fundamental laws of nature, suggesting a similar kind of view of emergence as a kind of logical consequence.

to know what will happen if we run a real-life auction using this or that set of rules, a good strategy is to construct game theoretic models of auctions and solve for the equilibria.

In the course of defending a form of "moderate realism" about psychology, Dennett argues that almost all of our discourse can be understood as justified in the same way that the use of models is. Just as it's often easiest to understand and predict a real-world system by constructing and analyzing a model of that system, it's often easiest to understand and predict a low-level system by first constructing a high-level, less fundamental model of it, and then analyzing that model.

Dennett gives a wide range of examples in which some system can be described in laborious and precise ways—giving the bit-by-bit description of some string of 1s and 0s, or the pixel-by-pixel description of a computer screen—but in which there are also available summary descriptions that are less accurate, but *much* easier to state and analyze. If I want to predict the next character in a string of binary digits, you can guarantee that I'll get the right answer by simply transmitting the whole string, but that might be a very large message. If the string is sufficiently "patterned," you'll be able to give me a very short summary description—e.g., "it's 70% 1s and 30% 0s, with no further discernible patterns"—that will let me do significantly better than chance.[19]

In Dennett's terminology, high-level descriptions of a system are justified when they identify "real patterns" in the system—when they convey information in a way that is both reasonably reliable, and substantially more efficient than giving the low-level, fundamental description of the system.[20]

David Wallace, who I'll quote at length, makes explicit that we can think of Dennett's view as providing a way to think about the relationship between non-fundamental and fundamental domains across the board—that he's not merely defending a view about psychology by offering an analogy to some examples from math and computer science, but is giving the seeds of an account of emergence in general:

> **Dennett's criterion.** A macro-object is a pattern, and the existence of a pattern as a real thing depends on the usefulness—in particular, the explanatory power and predictive reliability—of theories which admit that pattern in their ontology.

[19] While he doesn't call these summary descriptions "models," he does call them "idealizations," and it seems clear enough to me that he sees them as functioning just like models do.

[20] Really, we should probably think of this as a tradeoff—a summary description is justified to the extent that the sacrifice in reliability entailed by using it is compensated for by the gains in informational and computational efficiency.

> Dennett's own favourite example is worth describing briefly in order to show the ubiquity of this way of thinking: if I have a computer running a chess program, I can in principle predict its next move from analysing the electrical flow through its circuitry, but I have no chance of doing this in practice, and anyway it will give me virtually no understanding of that move. I can achieve a vastly more effective method of predictions if I know the program and am prepared to take the (very small) risk that it is not being correctly implemented by the computer, but even this method will be practically very difficult to use. One more vast improvement can be gained if I don't concern myself with the details of the program, but simply assume that whatever they are, they cause the computer to play good chess. Thus I move successively from a language of electrons and silicon chips, through one of program steps, to one of intentions, beliefs, plans and so forth—each time trading a small increase in risk of error for an enormous increase in predictive and explanatory power (pp. 50–1).

For Dennett, Wallace, and Carroll (whose views I'm about to discuss) there's a pragmatic element to non-fundamental ontology. The justification for describing reality in fundamental terms—whatever those turn out to be—is quite simple; that's just how the world is. But the justification for describing the world in non-fundamental terms—e.g., in chemical, or computational, or mental terms—is partly self-regarding; such descriptions are humanly tractable. Carroll (2016) encapsulates the view nicely:

> There is some fundamental description of our world, in terms of an evolving quantum wave function or perhaps something deeper. The other concepts we appeal to, such as "rooms" and "red," are part of vocabularies that provide useful approximate models for certain aspects of that underlying reality in an appropriate domain of applicability. (p. 352)

He explains the notion of usefulness he has in mind along the same broad lines as Dennett and Wallace, emphasizing the greater tractability of emergent levels of description:

> The reason why emergence is so helpful is that different theories are not created equal. Within its domain of applicability, the emergent fluid theory is enormously more computationally efficient than the microscopic molecular theory. It's easier to write down a few fluid variables than the states of all those molecules. (p. 99)

I have no novel arguments in defense of the Dennett/Wallace/Carroll view of emergence as modeling. Rather, I want to draw out some of its consequences for epistemology. I want to show that if you accept it, you should not think that the existence of a unique, maximally accurate, complete description of the world in fundamental terms—the existence of a final theory—entails that there exist similarly unique, canonical descriptions of the world in any set of non-fundamental terms. In particular, you shouldn't think there's a single, canonical accounting of who knows what, or who believes what, or who has what evidence. Rather, different epistemological models may be useful in different situations, without there being a canonical epistemological model to which they are all approximations.

Recall that if emergence were understood in terms of entailment, then it looked as if we *would* get canonical chemical, classical, economic, and mental descriptions of reality. So why are things different if we understand emergence in terms of modeling?

The key difference, already alluded to in the first Carroll quote above, is that some non-fundamental modeling framework may be able to provide a good, local model of *some aspect* of reality, without being able to provide a grand model of *all* of reality. Wallace is quite explicit on this point, though he uses slightly different terminology than I do:

> Crucially: this "reduction"... is a local affair: it is not that one theory is a limiting case of another *per se*, but that, *in a particular situation*, the "reducing" theory instantiates the "reduced" one. (p. 55, emphasis in original)

What he describes as one theory "instantiating" another, I would describe as one modeling framework being able to build a model that is a good model of another model built by another modeling framework. So in the terminology I'm using, an example of Wallace's point would be that while classical mechanics may be able to provide a good model of some fundamentally quantum mechanical system, there needn't be any general recipe for turning an arbitrary quantum mechanical model into its classical counterpart. And this is the example he goes on to give:

> The reason that classical mechanics is applicable to the planets of the Solar System is not because of some *general* result that classical mechanics is a limiting case of quantum mechanics. Rather, the particular system under consideration—the solar system—is such that some of its properties approximately instantiate a classical-mechanical dynamical system. Others do not,

> of course: it is not that the solar system is approximately classical, it is that it (or a certain subset of its degrees of freedom) instantiates an approximately classical system. (p. 56)

He goes on to point out that the very same quantum mechanical system—the solar system—might be approximated by various different classical models. We could treat it as nine point particles each undergoing mutual gravitational interactions, or we could simplify the model by ignoring the interactions between planets and only calculating the interactions between each planet and the sun, or we could complicate it by adding in all the asteroids, dwarf planets, and comets. Depending on our purposes and computational constraints, any of these classical models of the solar system might fit the bill. While there's a single, canonical answer to the question of how the solar system is fundamentally—albeit one that is probably beyond our capacities to grasp—there's no similarly unique, canonical answer to the question of how to model it classically.

If you accept the Dennett/Wallace/Carroll view, then I suggest that your default expectation should be that what goes for classical modeling of the solar system goes for modeling in non-fundamental terms more generally. Once we're aiming not to exactly limn the structure of reality, but to find tractable, useful models of it, we should be modest; we should expect that no one such model—even when we restrict our attention to models built using the same framework—will work best for all purposes.

Moreover, the farther we get from fundamental physics, the more powerful this argument becomes. Suppose our final theory of physics is couched in terms of quarks. Even so, it may be that there is a reasonably straightforward story to be told in terms of quarks, about what it takes to have a proton. And so one might reasonably hope that, even if the physicists' final theory won't mention protons, not *that much* modesty is called for when using modeling frameworks that *do* use protons, since the final theory will make it possible to offer a reasonably systematic and unified story about the existence and behavior of protons.

But things seem quite different when we imagine trying to get from the final theory to descriptions of the world in terms of economists' categories like *money*, or the *labor supply*, or even biologists' categories like *species*. Here, the distance between the non-fundamental categories and the fundamental ones seems vast. When there's that much mismatch between the categories we use in modeling the world, and the fundamental structure of the world, it seems reasonable to expect that the models we construct will be much less general and much more constrained in their applicability than when the categories we use fit the structure of the world more closely. This is why, even in the absence of some master argument for modest modeling in epistemology, it strikes me as a sensible starting point; beliefs are more akin to prices than to particles.

2.4 Methodological Upshots of Modesty

Titelbaum's distinction between modeling frameworks and models, introduced in the previous chapter, will let us distinguish two varieties of modest modeling:

Modesty in Framework Selection: A modest modeler may use different frameworks to model different phenomena, relying essentially on good judgment in making her selection, and without any hope of an uber-framework that would provide systematic rules for which framework to use when.

Modesty in Model Selection: Once the modest modeler has settled on a framework to use to model some phenomenon, the question of which model within that framework to construct may be similarly unsystematic. Perhaps the different models a framework allows the modeler to construct are in principle incapable of being merged into a single model, and good judgment is required to determine which models to use for which purposes.[21]

In the former case we have multiple frameworks, and our modeler makes a choice of which one to use. In the latter case our modeler has already settled on a framework, and must make a choice of which model within that framework to construct. Of course both sorts of modesty may be called for, even concerning the very same phenomenon—our modeler may make a non-algorithmic, judgment-guided choice about which framework to use, and *then* make a non-algorithmic, judgment-guided choice about which model to construct within that framework.

To take an example of modesty in framework selection, an economist might use a combination of judgment and heuristics to decide which framework for modeling growth to use in formulating her policy recommendations for an emerging economy. Should she use a framework that emphasizes human

[21] "Model Selection" has a more precise meaning in the statistical literature. See Ando (2010). Indeed, in the terminology of that literature, when one is engaged in "model selection" one generally *is* looking for systematic, algorithmic means for fitting a model to some body of data.

capital, or one that focuses on macroeconomic policy, or one that focuses on industrial policy, or instead one that stresses reducing corruption and strengthening property rights?[22] Our economist may have no hope of building an integrated framework that would let her construct models incorporating *all* factors that might legitimately go into a model of growth. Rather, she has an array of tractable modeling frameworks, and so must use experience-informed judgment to decide which ones to focus on in making her recommendations.[23]

What about modesty in model selection? Here we can use an example at the intersection of economics and philosophy. Suppose our modeler is set on using a Bayesian decision theoretic framework to model some phenomenon—her model will contain probabilistically coherent, expected-utility-maximizing agents. Even after settling on this framework, she's still faced with an array of choices. What states of the world should her model include? What actions should be available to the agents? What should their utility functions and probability functions be? Different settings for these parameters will generate different predictions about what the agents will do—or, if our modeler is doing normative modeling, prescriptions for what they *should* do—and if our modeler is modest, she may expect that the answers to these questions often cannot be answered algorithmically, but instead call for experience-informed judgment. We already saw in Rodrik's anecdote that the utility functions that work best for modeling people when they're trading commodities may not work best when modeling people in more personal interactions.[24]

How does modest modeling differ from ambitious modeling in practice? It's easy to suspect that it might not at all. If the modest modeler values explanatory power and scope—as I agree she should—then, like the ambitious modeler, she'll be happy when she can subsume what previously seemed like disparate phenomena under a single model.[25] And the ambitious modeler, just like the

[22] This example is from Rodrik (2015, p. 51).

[23] Of course she needn't pick just *one* framework. Often, when a modeler has multiple, incompatible frameworks to work with, she may check to see whether they agree in their predictions and/or recommendations. When they do, those predictions/recommendations are said to be "robust." See Levins (1966) for some classic discussion, focusing on tradeoffs between different modeling frameworks in population biology. But in cases where robustness is unavailable—where different models make very different predictions and/or recommendations, as in Rodrik's discussion of models of economic growth—researchers can't avoid exercising judgment in deciding which models to rely on.

[24] As an aside, at least some versions of the famous "problem of the priors" only arise for Bayesians whose ambitions are greater than those of a modest modeler. If you hope for a systematic, judgment-free recipe for using the Bayesian framework to construct models of particular situations, then you'll want systematic rules for assigning prior probabilities. But if, in general, you're comfortable with an ineliminable role for phronesis when it comes to how to apply the framework in particular cases, then the lack of a general solution to the "problem" of the priors is less likely to bother you.

[25] That is, the modest modeler just as much as the ambitious one can regard unification, in someting like the sense described by Friedman (1974), as an explanatory virtue.

modest modeler, will in practice have to settle for partial and less than fully accurate models. So it can seem like both ambitious and modest modelers may aim to construct exactly the same sort of models with exactly the same sorts of virtues, differing only in their private hunches about just how much room for improvement exists. But why should that difference in private hunches make a methodological difference?[26]

Ultimately, I think it's not true that modest and ambitious modelers will construct exactly the same sorts of models; rather, they'll often differ in their judgments about which research programs are promising, which might lead them to focus on different projects. I claim we often find ourselves in the following sort of situation. A given modeling framework can account for some range of phenomena, but only by subsuming the phenomena within multiple, incompatible models. The ambitious modeler finds this unsatisfying; she'll be on the lookout for an alternative framework that could, in principle, account for all of the phenomena in a single model. Even if she doesn't expect that we'll ever be able to fully describe that model, getting clearer about some of the features it would need to have will seem like a promising, worthwhile research program. By contrast, because the modest modeler is agnostic about whether there's room for improvement over the given modeling framework, she'll see less motivation to explore alternatives, at least before she's been given some positive reason to think that superior alternatives exist.

In fact, I think this is the situation with respect to a wide range of debates in contemporary epistemology. In the remainder of this book I'll consider particular debates that seem to me to look quite different once we view them through the lens of modest modeling. Before jumping in, it will help to give a kind of argument schema that I'll return to again and again, with the details differing from chapter to chapter.

First, and largely implicitly, I'll interpret views about phenomena of epistemological interest—information, belief, knowledge—as modeling frameworks. Chapter 3 will consider the possible worlds framework for modeling information, Chapter 4 will consider the Bayesian framework for modeling belief and learning, Chapter 5 will consider the relationship between the modeling frameworks of decision theory and folk psychology, and the next two chapters will consider modeling frameworks that allow for trivial iteration of higher order belief and knowledge (Chapter 6) as well as the possibility of common knowledge (Chapter 7). And in keeping with the Kuhnian methodological remarks of §1.2, I'll interpret objections that are ordinarily

[26] Thanks to an anonymous referee for raising this concern.

intended to show that some theory is false—that some general principle has a counterexample—as attempts to show that some modeling framework cannot adequately capture its target phenomenon.

Second, I'll argue that these objections can be met—the phenomena *can* be captured—so long as we're willing to be modest about model selection. The objections show that no *single* model built by the framework in question can capture all the cases we want to. But we can capture the phenomena the objections point to by using different models to handle different cases. In each debate I'll discuss, an even *more* modest/pluralist take than mine would hold that the phenomena are best captured not by multiple models within a single framework, but by multiple frameworks; i.e., the phenomena call for modesty in framework selection.[27] For instance, a live option in recent debates about belief and credence—a topic I'll discuss in Chapter 5—is known as "dualism" (Jackson, 2019), and I'll suggest it's best understood as the view that folk psychology and decision theory are distinct and ineliminable modeling frameworks for understanding human behavior—both are necessary, and neither can do the work of the other. By contrast, as will come out in that chapter, I'm agnostic, leaning skeptical, about whether folk psychology can do explanatory work over and above what's done by decision theory. So I don't think there's any strong default rule to the effect that the maximally modest take on some debate is always the right one.

I'll admit that the objections I'll consider point to areas in which alternative modeling frameworks could *conceivably* outperform the frameworks I'll be defending; if alternative frameworks could build comparably simple, tractable models, but with more explanatory power and scope than those of the frameworks I'll defend, then that would be a reason to discard the frameworks I'll discuss in favor of the alternatives. But I'll suggest that—at least in the particular debates I'll be discussing—this isn't the state of play. And moreover, we shouldn't hold our breath—the categories used by the frameworks I'll be discussing are far enough from the fundamental that we shouldn't expect that *any* modeling framework using those categories will be able to build very simple models with very broad scope.

[27] This isn't always a crisp distinction; deciding when models are similar enough to count as belonging to a single framework can be a delicate matter.

3
Modeling with Possible Worlds

In this chapter I'll defend the possible worlds framework for modeling the contents of belief. Both the threats against which I'll defend it—the problems of coarse grain—and the "fragmentationist" response I'll offer are familiar. But, at least as a sociological matter, the fragmentationist response has been unpersuasive, I suspect because it can look like an ad hoc patch—an unmotivated epicycle aimed at saving a flailing theory from decisive refutation. I'll argue that in fact, a fragmentationist version of the possible worlds framework is exactly what we should have expected all along, *if* we are modest modelers. So my aim in this chapter is twofold. On the one hand, I want to make a contribution to the first-order philosophical debate about how to model propositional attitudes like belief. But I also want to use that debate to illustrate the fruitfulness of the modest modeling approach.

3.1 Possibilities

The foundational insight of the possible worlds framework for thinking about the content of belief and knowledge—really, for thinking about *information* in general, though I'll focus on belief and knowledge here—is that we can understand the content of someone's beliefs in terms of what possibilities those beliefs rule out.[1] To take an example of Daniel Dennett's, suppose a toddler says "my daddy is a doctor."[2] As the child grows, so does his understanding of what this sentence means. Perhaps initially he can't elaborate at all when you ask him what a doctor is. Later, he can explain that doctors help sick people. Still later, he'll learn about differences between doctors, EMTs, therapists, and so on. When should we say the child genuinely understands that his daddy is a doctor? Dennett's point in introducing this example was to make trouble for

[1] For many applications, we'll also need a probability distribution over those possibilities. More on this later.

[2] The example is originally from Dennett (1968), though it is developed by Stalnaker (1984) specifically in the context of motivating the possible worlds approach to belief.

Idealization in Epistemology: A Modest Modeling Approach. Daniel Greco, Oxford University Press.
DOI: 10.1093/oso/9780198860556.003.0004

views on which the content of our entire body of knowledge can be happily divided into individual, sentence-like items of knowledge.[3]

But the example has also been used to make a more positive point; while gradual increases in understanding are difficult to model if we think of bodies of information as composed of individual, sentence-like items of information, it's much easier if we think of information in terms of ruling out possibilities:

> The general point that Dennett is making is that the capacities and dispositions that constitute a person's knowledge do not divide up neatly into discrete items of information—into propositions. But one of the virtues of the possible worlds analysis is that it does not require such a division. One can describe very naturally the situation in which one is tempted to say that a person only "sort of" knows something, and one can explain very naturally how a person's understanding of a proposition can grow. The child who says his daddy is a doctor understands what he says, and knows it is true, to some extent, because he can divide a certain, perhaps limited, range of alternative possibilities in the right way and locate the actual world on the right side of the line he draws. As his understanding of what it is for Daddy to be a doctor grows, his capacity will extend to a larger set of alternative possibilities or, rather, the extension of this capacity will be the growth of his understanding. (Stalnaker, 1984, p. 64)

While within philosophy the possible worlds framework is often associated with Lewis and Stalnaker—though there are many precursors[4]—essentially the same approach is found in other disciplines that study information as such. In addition to the examples of linguistics and economics, there's a whole branch of computer science called "information theory," in which thinking of information in terms of ruling possibilities out plays a foundational role. The basic insight of information theory is that information can be quantified; we can talk rigorously about messages carrying more or less information. And information theorists understand the informativeness of a message in terms of the measure of the set of possibilities it rules out—the more possibilities a message rules out, the more informative it is.[5]

[3] That is, he was defending a "holist" view in the philosophy of mind. See Fodor and Lepore (1992) for a survey sympathetic to the "atomist" position, opposed to holism.

[4] Leibniz makes use of the notion of possible worlds throughout his writings, and puts them to the same sort of work that contemporary metaphysicians do. See also Wittgenstein (1921) and Carnap (1967).

[5] The founding text of the field is Shannon (1948).

Note that while I've described a framework in which we think about having information as distinguishing between possibilities, I haven't said anything about what possibilities themselves are. While Lewis (1986) had an ambitious and systematic theory about this, we needn't endorse any such theory in order to be sympathetic to the possible worlds approach to information. In fact, I think a modest modeler can probably decline to offer a general theory of the metaphysics of possibilities, without embarrassment,[6] and I think the arguments I'll make in this chapter are compatible with a variety of views about the nature of possibilities.

Before moving on, I want to make a bit clearer just how ecumenical I intend the approach discussed in this chapter to be. In various writings, Chalmers (2011a, b) defends a version of the possibility-based approach to modeling content. In his framework, the possibilities in question are epistemic possibilities of a certain sort—they're possibilities that cannot be ruled out a priori by an ideally rational agent. But they needn't be metaphysical possibilities—if an ideally rational agent can't rule out a priori that water is not H_2O, then Chalmers' framework will happily make use of possibilities in which water is not H_2O to represent the knowledge of chemically unsophisticated subjects. Stalnaker (1984), by contrast, can't avail himself of such possibilities. To oversimplify greatly, he gives causal facts an important place in solving the problem of intentionality, and because water and H_2O have identical causal powers, he can't easily allow that we can have beliefs or knowledge about the one without having beliefs or knowledge about the other. Nothing I'll say in this chapter will commit me one way or the other on this debate, so I use the phrases "possible worlds framework" and its cousins with some trepidation, as it might misleadingly suggest siding with Stalnaker over Chalmers.

3.1.1 Problems of Coarse Grain

This is not the place to survey the virtues of the possible worlds approach to thinking about information—that would take many books, and they've already been written. Why, then, is it out of favor among philosophers? While the reasons are varied, I think it's fair to say that they mainly stem from the thought that if information is understood in terms of ruling out possibilities, the resulting picture is overly "coarse-grained"—sometimes we

[6] See Stalnaker (1996) for some relevant discussion.

want to distinguish between two bodies of information, even when they rule out the same possibilities.[7]

Perhaps the most obvious manifestation of this problem concerns necessary truths—claims that are true no matter which possibility is actual. If we're using the possible worlds framework to model an agent's knowledge, then any claim true in all the worlds in our model will be one the agent counts as knowing; in the framework, an agent only fails to know some claim if there are worlds she can't rule out in which the claim is true. But this looks to make it too easy to have knowledge of necessary truths. Should we really be committed to modeling every agent as knowing the disjunction P ∨ ~P for any arbitrary P, just because we don't include worlds in our models where P ∨ ~P is false? This challenge is known as the "problem of logical omniscience."

A quick response might point out that, as long as we're not yet committal on just what possibilities are—e.g., if we haven't endorsed Lewis' modal realism—it's not obvious that there's any barrier to including worlds in our models where P ∨ ~P is false.[8] So why not hold onto the possible worlds framework for modeling information, while including worlds in our models where logical truths are false, when we want to model logically ignorant agents? While this "impossible worlds" approach has been defended, in my view it ultimately has to give up on much of the power of the possible worlds framework.[9] Part of what makes the possible worlds framework so fruitful is that it can draw on the resources of set theory to model relations between bodies of information—this allows us to represent the information carried by logically complex sentences as determined by the information carried by their parts. Whatever set of possibilities corresponds to some sentence P, we can think of ~P as corresponding to the complement of that set. And whatever set of possibilities corresponds to Q, we can think of P ∨ Q as corresponding to the union of the sets corresponding to P and Q. But once we do this, even without any specific commitments about the nature of the possibilities among which we're distinguishing, we get the result that P ∨ ~P will always correspond to the entire set of possibilities; if you start with some set Ω—whether it's conceived as a set of concrete worlds, or sentences, or anything else—and then take some subset s of Ω, the union of s and its complement in Ω is Ω. So there's no easy route to avoiding the consequence that, in the possible worlds framework, agents are always modeled as knowing all truth functional tautologies.

[7] See Field (1978) and Soames (1987).

[8] This move isn't available to Chalmers, insofar as he thinks that ideally rational agents are logically omniscient, and so epistemic space doesn't include any worlds where logical truths fail to hold.

[9] Though see Berto and Jago (2018) for a sympathetic survey.

A mirror image of this problem concerns claims that are not necessarily true, but necessarily false. Sometimes our beliefs are inconsistent—they can't all be true. But this looks difficult to capture in the possible worlds framework; if my body of belief is inconsistent, then it rules out *every* possibility. If we're looking for a set of possibilities corresponding to that body of belief, the empty set seems to be the only option. But then we get the unattractive result that there is only one inconsistent body of information: ∅. Suppose you believe that all dogs go to heaven but Fido the dog does not go to heaven, while I believe all mammals have hair but whales do not have hair. While both of our bodies of belief are inconsistent, they are not *identical*. And yet, if we're trying to use the possible worlds framework to model the content of these bodies of belief, ∅ seems to be the only option in both cases. We can call this the "problem of inconsistency."

Lastly, there are difficulties that involve neither necessity nor impossibility. Suppose I believe P, and P ⊃ Q, both of which are contingent. In the possible worlds framework, we'll model this by representing my body of belief with a set of possibilities, in all of which both P, and P ⊃ Q, are true. But in that case, we've thereby modeled me as also believing that Q—if all the possibilities compatible with my beliefs are possibilities in which both P and P ⊃ Q are true, then all the possibilities compatible with my beliefs are ones in which Q is true. But isn't it possible that I've failed to make the leap, and haven't actually inferred Q? Shouldn't we want to be able to represent the beliefs of an agent who hasn't drawn out all the consequences of her beliefs? While this is sometimes itself called the problem of logical omniscience, to distinguish it from the first problem I mentioned, let's call it the "problem of deduction."[10]

One common response to each of these problems is to start talking about idealization. Perhaps the possible worlds framework is really a framework for modeling the beliefs of an ideal agent, and this sort of agent never has inconsistent beliefs, always believes the consequences of her beliefs, and (as a special case of the former) knows all logical truths. Or perhaps the framework captures what non-ideal agents like us *would ideally* believe, given our evidence.[11] While I have some sympathy for these sorts of response, I think they're ultimately too concessive. They invite the thought that, whatever the fruitfulness of the

[10] "The Problem of Deduction" is what (Stalnaker, 1984, ch. 4) calls the general family of problems that I am calling problems of coarse grain.

[11] See, e.g., Colyvan (2013). My approach here is more in line with Yap (2014), who does see these models as involving idealization, but not distinctively *normative* idealization. Like Yap, I don't think we need to think of these models as aimed primarily at capturing a distinctive, rationally or normatively ideal sort of agent. See also Appiah (2017), who defends an approach to idealization in decision theory to which I'm very sympathetic, and which this chapter is largely consonant with.

possible worlds framework as an approach to modeling ideal agents, we should be on the lookout for a different approach to modeling the beliefs of non-ideal agents like us. And ultimately I want to resist that thought. The difference between ideal and non-ideal agents, on the picture that will emerge, isn't that entirely different frameworks are appropriate for modeling the beliefs of one and the beliefs of the other. Rather, the difference is that the information of an ideal agent can be perfectly captured by a single model in the possible worlds framework, while to model the beliefs of non-ideal agents, we'll need to be more flexible, sometimes using one model, sometimes another.[12] For an ideal agent, the question of how to use the possible worlds framework to represent her beliefs admits of a fixed, context-independent answer, while for non-ideal agents like us, it does not; just which model we should construct to represent the beliefs of a non-ideal agent depends on our purposes, and different models will be appropriate to different situations. But if we're already modest modelers, this shouldn't seem like much of a surprise—this is typical of what's required when using neat models to represent a messy reality.

3.2 The Fragmentation Response

The response pursued independently by Lewis and Stalnaker is to posit that our belief states are "fragmented." In a nutshell, rather than modeling an agent's body of information with a single set of possibilities, we model them with multiple sets. In this section I'll offer a sketch of how this idea can be used to respond to the problems of coarse grain.

Lewis (1982) gives a canonical statement of the "fragmentationist" position approach to the problem of inconsistency:

> I used to think that Nassau Street ran roughly east-west; that the railroad nearby ran roughly north-south; and that the two were roughly parallel. (By "roughly" I mean "to within 20°.") So each sentence in an inconsistent triple was true according to my beliefs, but not everything was true according to my beliefs. Now, what about the blatantly inconsistent conjunction of the three sentences? I say that it was not true according to my beliefs. My system of beliefs was broken into (overlapping) fragments. Different

[12] As will come out in the next chapter, this is a bit of an oversimplification. If we want to capture the beliefs of an agent not just at a single time, but over time, then even an ideally rational agent might not be representable with a single possible worlds model, for reasons that will come out in §4.3.

> fragments came into action in different situations, and the whole system of beliefs never manifested itself all at once. The first and second sentences in the inconsistent triple belonged to-were true according to-different fragments; the third belonged to both. The inconsistent conjunction of all three did not belong to, was in no way implied by, and was not true according to, any one fragment. That is why it was not true according to my system of beliefs taken as a whole. (p. 436)

In the paper from which this quote is drawn Lewis doesn't mention possible worlds (uncharacteristically). But it's easy enough to see how the fragmentationist move can be used in the possible worlds modeling framework. If you want to use possible worlds to model Lewis' beliefs about the layout of Princeton, you shouldn't just use the empty set; that would leave you unable to distinguish Lewis' beliefs from some completely unrelated body of inconsistent beliefs, and would also have the unhappy consequence that none of his actions—e.g., walking perpendicular to Nassau street in the hopes of reaching the railroad—could be informatively explained by appeal to his beliefs. Instead, you'll need two sets of possibilities—one set of possibilities throughout which both Nassau street and the railroad run east-west, and another set throughout which they both run north-south.

The approach to the problem of deduction is, essentially, the same. Imagine a slight variant on the example where Lewis' beliefs aren't inconsistent, but they fail to be deductively closed. Suppose he believes that Nassau street runs east-west, and that the railroad runs east-west, but fails to believe that Nassau street and the railroad are parallel. As before, we can't model this body of belief with a single set of possibilities, but we can do with two. In one set, Nassau street runs east-west throughout the space of possibilities, but the possibilities differ on the orientation of the railroad. In the other set, the situation is reversed—the railroad runs east-west throughout the space, but they differ on the orientation of Nassau street. So it will be true in a fragment of Lewis' belief state that Nassau street runs east-west, and true in a (different) fragment that the railroad runs east-west, but true in no fragment that Nassau street and the railroad are parallel.

What about the problem of logical omniscience? Here, while I think fragmentation has a role to play, it's only one ingredient of the proper reply. The primary ingredient will be a "contingent surrogate" for the putatively necessary truth that we want to model an agent as ignorant of. That is, the possible worlds framework has an easy time explaining how we can be ignorant of contingent truths, so a natural strategy for handling ignorance of putatively necessary information, is to look for a contingent surrogate for that information; what you were *really* ignorant of wasn't the necessary truth N, but some contingent truth T. Stalnaker (1984) famously suggested that, at least sometimes, this contingent surrogate will be a metalinguistic claim:

> For example, one who has not yet figured out whether a formula such as "((P ⊃ ((Q ⊃ P) ⊃ Q)) ⊃ (P ⊃ Q))" is a tautology is like one who has not yet processed some syntactically complex sentence of a natural language. The information which he lacks seems, intuitively, to be metalinguistic information—information about the formula. (p. 74)

Since metalinguistic information is not necessary—facts about which sentences correspond to which possibilities depend on contingent facts about how language is used—we can easily model agents as lacking metalinguistic information within the possible worlds framework. But as he goes on to acknowledge, in many cases, when we seem to be ignorant of necessary truths, it's not particularly natural to describe us as lacking metalinguistic information. Rayo (2013a, ch. 4) describes a case in which someone is building a fence to enclose a patch of land, and while she knows the shape of the land (square), and its area—81 square meters—she doesn't know how much fencing she needs to buy. Here, it's natural to describe the agent as ignorant of the necessary truth that $\sqrt{81} = 9$, but while there's a ready contingent surrogate to hand, it's not metalinguistic. The information she lacks is isn't information about numerals, but instead the information that 36 meters of fence will suffice to enclose her patch of land.

We're not home free. Rayo, responding to a challenge raised by Field (1978), explains that coming up with a contingent surrogate often isn't enough to satisfactorily deal with the problem of logical omniscience. Field points out that an agent who is ignorant that some complex formula expresses a tautology will typically have *other* contingent metalinguistic information—e.g., information about how the symbol "⊃" is used—from which it follows that the formula does, in fact, express a tautology. And in Rayo's example, I stipulated that the agent in question knows that she has a square piece of land whose area is 81 square meters. And all the possibilities in which she has a square piece of land whose area is 81 square meters are possibilities in which she has a piece of land that can be enclosed by 36 meters of fence. In short, even once we come up with a contingent surrogate, we run back into the problem of deduction.

But we've already seen that we can address the problem of deduction via fragmentation. So to deal with an instance of the problem of logical omniscience—to model a case in which it's initially tempting to describe an agent as ignorant of some necessary truth N—we first find a contingent surrogate for N, C, such that we can model the agent as ignorant of C; her body of beliefs will be modeled by a set of possibilities that includes both C possibilities, and ~C possibilities. Having done that, we'll often run back into the problem of deduction, since the agent will already have beliefs that entail C. In metalinguistic cases, these might be beliefs about basic rules of use for the words in some language. In Rayo's case, this is the agent's belief that she has a square-shaped, 81 square meter patch of land. So the second step is to invoke fragmentation. The agent in question has the information that C in some fragment(s) of her overall belief state, but not all. If we are to model Rayo's agent, we will need at least two sets of possibilities, in one of which it is true throughout the set that the agent has a square-shaped, 81 square meter patch of land that (ipso facto) can be enclosed by a fence 36 meters long. But in the other set of possibilities, it's not true throughout the set that her land has the aforementioned dimensions.

Phew. Unless you've already drunk the possible worlds kool-aid, I imagine the last few pages look like a series of increasingly awkward contortions aimed at fitting some obviously recalcitrant phenomena into a woefully inadequate modeling framework. My aim in the rest of this chapter will be to disabuse you of that notion. In a nutshell, I'll argue as follows. First, while the problems of coarse grain may seem to specifically target the possible worlds framework, they are in fact quite general, and arise in one form or another for any view on which beliefs and values combine to rationalize and explain actions. And second, I'll show how we can see the fragmentationist response as a special case of modest modeling. So neither the difficulties to which the possible worlds theorist is responding nor the general shape of her response are parochial or ad hoc.

3.3 Why Fragmentation Is Not Ad Hoc

3.3.1 Belief and Action

So far I've talked about how the possible worlds framework lets us model the content of a body of belief, but I haven't yet said anything about when an agent is fruitfully modeled as believing some particular content; it's one thing to have a story about how to model bodies of information, and another to have a story about which agents should be modeled as believing which bodies of information.

In this section I'll argue that the problems of coarse grain all arise from a general commitment concerning what it takes for an agent to believe some body of information: an agent's beliefs must be poised to guide her actions. If we adopt this view of belief, then sooner or later we'll run into the problems of coarse grain. But the idea that belief is essentially action guiding is much more attractive and widely held than is a commitment to the possible worlds framework. So the problems of coarse grain are (almost) everyone's problems.[13]

Before moving on, I want to offer another clarificatory note concerning the limits of my ambitions in this chapter. There are philosophers who deny any tight metaphysical connection between belief and action. Galen Strawson (1994) famously described a species of "weather watchers" who have beliefs about the weather, but have no desires about anything and are constitutively incapable of action. More generally, adherents of the phenomenal intentionality research program (Kriegel and Horgan, 2013) tend to hold that the content of a subject's beliefs depends only on her phenomenology, and not at all on her dispositions to act. While I myself am doubtful that the notion of phenomenology is sturdy enough to bear so much philosophical weight,[14] for present purposes I don't think I need to take a stand on this question. Even phenomenal intentionalists should allow, in a quasi-stipulative sense, a notion of belief-as-the-basis-of-action. For example, Declan Smithies (2019a) is primarily interested in a notion of belief that is closely connected to phenomenology, but holds that we can also talk about representational states whose constitutive role is to guide action, and that such states play an ineliminable role in cognitive science. So throughout this chapter, if I ever get ahead of myself and talk about *the* role of belief in guiding action, please interpret me as talking about *a* role of belief. I'll argue that if you want a notion of belief that plays this action-guiding role, you'll end up with something like the problems of coarse grain. But this is consistent with allowing that there are other notions of belief that play other roles, and which don't face these problems.

This strategy is consilient with my remarks in Chapter 1 about rejecting modeling frameworks. My claim in this chapter is that the possible worlds modeling framework has virtues that we shouldn't expect some rival

[13] This is in the spirit of Stalnaker (1991), who argues that the problem of logical omniscience is "a symptom of a tension in our ordinary conceptions of knowledge and belief."

[14] I'm sympathetic to the criticisms in Dennett (1988).

framework to replicate. But that's consistent with conceding that it can't do everything we might initially have wanted a theory of belief to do; for all I say in this chapter, we may need other modeling frameworks to capture other aspects of belief.

William James said that "the test of belief is willingness to act."[15] But what is it to act on a belief? A natural answer: it's putting your money where your mouth is, acting in ways that would lead to good results if the belief were true, but not otherwise. While this is a fine start, it skirts over complexities. Suppose I believe it's going to rain. What is it to act on that belief? A natural answer: to carry an umbrella, or wear a raincoat, or stay inside. But, as Geach (1957) pointed out long ago, this won't quite do. What if I don't mind or even enjoy getting wet? In that case, acting on my belief that it will rain won't involve any of those behaviors.

The typical lesson philosophers have drawn from this observation is that while there is a connection between what someone believes, and how she behaves, we can't associate individual, sentence-like beliefs with dispositions to act. Rather, the association must be holistic; only large systems of both belief and desire can be associated with particular actions. Instead of saying: "if you believe P, you'll do ϕ, while if you believe Q, you'll do ψ..." we say: "if you believe P, and Q, and R, and S... and you desire D, and E, and F, and G... then you'll do ϕ."

Stalnaker offers a version of this idea:

> To desire that P is to be disposed to act in ways that would tend to bring it about that P in a world in which one's beliefs, whatever they are, were true. To believe that P is to be disposed to act in ways that would tend to satisfy one's desires, whatever they are, in a world in which P (together with one's other beliefs) were true. (1984, p. 15)

Better. Still oversimplified—in particular, it says nothing about degrees of belief or desire—but enough of an improvement that we can draw some interesting philosophical lessons from it. In particular, this holistic understanding of the relationship between belief, desire, and action is enough to lead to the problems of coarse grain. And natural elaborations of the idea only make the problems worse.

Let's start with the problem of inconsistency. Suppose an agent believes both P and ~P. Given the Stalnakerian account of belief above, this means that the

[15] See Zimmerman (2018) for a recent, sympathetic exposition of this Jamesian idea.

agent is disposed to act in ways that would tend to bring about what she desires, in a world in which P and ~P (along with the rest of her beliefs) were true. But there are no worlds where P and ~P are true, so it's hard to see this as placing a substantive constraint on how one must be disposed to act, if one has inconsistent beliefs.

To make this a bit more concrete, let's return to the example Lewis discusses. Lewis believes that Nassau Street runs east-west, the railroad runs north-south, and that Nassau Street and the railroad are parallel. Suppose Lewis is walking on Nassau Street, and wants to reach the railroad. What should we expect him to do, given his beliefs? In particular, should we expect him to keep walking straight, or to turn and walk perpendicular to Nassau street? Well, if it really were true that Nassau Street ran east-west, the railroad ran north-south, and the two streets were parallel, then what would it take to get from Nassau Street to the railroad?

That question brings us into the realm of "counterpossibles"—subjunctive conditionals with impossible antecedents. An influential view on these matters says that any conditional of the form "if it were the case that P, then Q," where P is impossible, is true.[16] *If* that view is right, then we run into a very similar problem of inconsistency to the one that arose for the possible worlds framework for modeling the contents of belief. That is, *any* action ϕ will be such that, if some impossibility were true, then doing ϕ would tend to bring about the satisfaction of your desires. So once we attribute inconsistent beliefs to an agent, the holistic belief-desire-action connection adverted to above no longer places any substantive constraints on how the agent will act. Just as the possible worlds framework (without fragmentation) can't distinguish between different bodies of inconsistent belief, the holistic belief-desire-action connection can't distinguish between the behavioral dispositions of agents with (different bodies of) inconsistent beliefs.

I'll return shortly to the question of whether we should accept that counterpossibles are always trivially true. First, I want to show that the problems of deduction and logical omniscience also arise for the holistic belief-desire-action connection, via the connection with principles concerning the logic of counterfactuals.

Next, the problem of deduction. Suppose Lewis believes that Nassau street runs east-west, and that the railroad runs east-west, but doesn't believe that they're parallel. Again, let's assume he's walking on Nassau street and wants to reach the railroad. How should we expect him to act? The most natural

[16] See, e.g., Williamson (2007, ch. 5).

way of applying the holistic belief-desire-action principle would lead us to assume that he'll turn perpendicular to Nassau street; if it were true that Nassau street runs east-west, and true that the railroad runs east-west, then a good way of satisfying his desire of getting to the railroad from Nassau Street would involve turning perpendicular to Nassau street. But this doesn't seem to capture his failure to believe that Nassau Street and the railroad are parallel. Why not? First, assume that the "substitution of equivalent antecedents" (SEA) in counterfactuals is valid.[17] That is, for any conditional of the form "if it were the case that P, then it would be the case that Q," we can substitute some other antecedent logically equivalent with P, without changing the truth value of the conditional as a whole. If that's right, then the holistic belief-desire-action connection will invite us to predict that agents whose beliefs aren't dedeductively closed will tend to act exactly like agents whose beliefs are deductively closed. That is, drawing out all the consequences of your beliefs would make no difference to how you would act. That seems crazy—why would we be under any pressure to say it? Consider the following two conditionals:

1. If Nassau Street ran east-west and the railroad ran east-west, then turning perpendicular to Nassau Street would be a good way to get to the railroad.
2. If Nassau Street ran east-west and the railroad ran east-west and Nassau Street and the railroad were parallel, then turning perpendicular to Nassau street would be a good way to get to the railroad.

If SEA is valid, then these two conditionals must have the same truth value, because the antecedent of (2) is equivelent to the antecedent of (1). More generally, for any deductively open body of belief P, there is a deductively closed body of belief P' that is equivalent to P, which we can obtain just by starting with P and adding the consequences that weren't already included in it. By SEA, a conditional "if it were the case that P, then performing actions ϕ would satisfy one's desire that Q" will have the same truth value as "if it were the case that P', then performing actions ϕ would satisfy one's desire that Q." So given the holistic belief-desire-action connection, while we're not exactly barred from interpreting agents as having deductively open bodies of belief, the distinction between deductively open and deductively closed bodies of belief looks to make no difference, when it comes to action. And of course, this is deeply implausible—intuitively, we'd expect that the version of Lewis who had realized that the streets were parallel would be much more likely to turn

[17] See Warmbrōd (1981) for a canonical defense.

perpendicular to Nassau street than the one who only believed both streets ran east-west, but hadn't yet inferred that they were parallel. And remember, we got to this version of the problem of deduction just by assuming that agents will tend to act in the ways that would satisfy their desires, if their beliefs were true, together with a popular principle about the logic of counterfactuals. We did not go via a possible worlds analysis of contents.

The problem of logical omniscience for the holistic belief-desire-action connection falls out as a special case of the problem of deduction. Just as we can't behaviorally distinguish between the agent who's inferred that the two streets are parallel from the agent who merely believes that they both run east-west, we can't distinguish the agent who's inferred any logical truth you like from an agent who hasn't. If doing ϕ would tend to satisfy your desires, if your beliefs were true, then doing ϕ would *also* tend to satisfy your desires, if your beliefs *and* some complex tautology were true. So endorsing the holistic belief-desire-action connection leaves us unable to distinguish those agents who believe complex tautologies from those who don't.

It might be objected that I haven't *really* shown that the problems of coarse grain arise independently of the possible worlds framework for modeling content. After all, the principles of the logic of counterfactuals I appealed to in deriving the problems of coarse grain from the holistic belief-desire-action connection—namely, that counterpossibles are trivially true, and that SEA is valid—derive much of their support from possible worlds analyses of counterfactuals.[18] So perhaps it's no surprise that the same problems that arise for possible worlds models of content *also* arise for possible worlds analyses of counterfactuals.

I'll offer a couple short responses, along with a more substantial one. First, *even if* I need to assume a possible worlds analysis of counterfactuals, that doesn't make the arguments above uninterestingly question-begging. It's sometimes assumed that hyperintensional analyses—ones that, unlike possible worlds analyses, make distinctions between necessarily equivalent propositions—are appropriate when we're talking about representation, but not when doing serious metaphysics.[19] Limited creatures like us distinguish between water and H_20, Hesperus and Phosphorus, and $2 + 2 = 4$ and Fermat's last theorem, but objective reality does not. Suppose you accept this way of carving up theoretical space, and you place questions about counterfactuals

[18] The defense I mentioned of SEA, due to Warmbrōd (1981), was couched within a possible worlds framework.

[19] See Nolan (2014), who doesn't himself endorse this distinction, but finds it in others.

on the reality side of the reality/representation distinction and questions about belief on the other. In that case, you might have thought that belief should be modeled in some hyperintensional, not-merely-possible worlds framework, but counterfactuals can be given a possible worlds treatment. The arguments I've just offered put pressure on that combination of views. If you accept the holistic belief-desire-action connection, and you accept the logical principles governing counterfactuals that happen to fall out of a possible worlds treatment, then you'll run right into essentially the same problems that face possible worlds analyses of belief.

Second short response. While I'm not persuaded that there are counterexamples to the triviality of counterpossibles, or SEA—see Williamson (2018) for a powerful recent defense of semantic orthodoxy against a variety of objections—suppose that's wrong, and these general principles fail. It's certainly true that there are some intuitively compelling examples of non-trivial counterpossibles, e.g.:

Solid: If there was a piece of steel in the shape of a 36 sided platonic solid, it would have more sides than any piece of steel in the shape of a dodecahedron.[20]

Solid looks like a non-trivially true counterpossible—non-trivial because the same intuition that says **Solid** is true says that it would be false if we replaced "more" with "fewer" in the consequent. But even if that's right—though again, see Williamson (2018) for an error theory that explains away our inclination to think it is—it does not suggest a strategy for making non-trivial sense of the counterpossibles that, together with the holistic belief-desire-action connection, led to the problems of coarse grain for belief. Even if you think **Solid** is non-trivially true, I don't see why you should think the question below has a non-trivial answer:

How to Navigate? If Nassau Street runs east-west, the railroad runs north-south, and Nassau Street and the railroad are parallel, how should one get from Nassau Street to the railroad?

But given the holistic belief-desire-action connection, we'd need a non-trivial answer to **How to Navigate?** in order to make sense of Lewis' inconsistent beliefs. That is, only if **How to Navigate?** has a non-trivial answer can we attribute to Lewis the particular inconsistent beliefs he's supposed to have

[20] The example is from Nolan (2014).

on the basis of his dispositions to act. Along similar lines, even if there are *some* counterexamples to SEA, the pair of counterfactuals 1 and 2 from earlier in this section don't look like one. But they would need to be a counterexample for us to make sense of a distinction between a deductively open and a deductively closed body of belief in my modified version of Lewis' example.

In short, it's one thing to argue against general principles of the logic of counterfactuals like the triviality of counterpossibles, and SEA. It's another thing to argue that the instances of those principles that lead to the problems of coarse grain are themselves counterexamples to those principles. Even if we reject the principles in full generality, the particular instances that we need in order to generate the problems of coarse grain look comparatively innocuous.

While I said earlier that I would try to be as neutral as possible about how to understand the nature of possibilities, I do think the preceding considerations suggest certain lessons. If we want a holistic belief-desire-action connection, and we want to model beliefs using sets of possibilities, then whatever possibilities we use will have to have the following feature. We'll have to be able to make non-trivial distinctions concerning which actions would further or frustrate an agent's desires, if those possibilities were actual. Depending on our views about the nature of metaphysical possibility, this may or may not amount to letting in metaphysical impossibilities. But not just any "possibilities" will do; when possibilities get weird enough—in particular, when they get so weird that our ability to make sense of which actions would further which goals if those possibilities were realized breaks down—then they become unsuited to model belief, if belief is to fit into a holistic belief-desire-action connection.[21]

My more substantial response to the idea that I've merely defended one part of the possible worlds framework (the account of information) by appeal to another (the account of counterfactuals) is to argue that appealing to claims about counterfactuals isn't really essential to derive problems of coarse grain from the holistic belief-desire-action connection. The version of that connection I've been appealing to so far is one that makes use of a counterfactual—it talks about what actions *would* make an agent's desires come true, if her beliefs were true. But as I alluded to when I introduced that formulation, this presupposes what is, for many purposes, an oversimplified framework for

[21] I myself am sympathetic to the view of metaphysical possibility defended by Rayo (2013a), on which the limits of metaphysical possibility are tightly linked to the limits of intelligibility. So I think it's only once possibilities are "weird enough" that we should regard them as metaphysically impossible. But if we accept that there are metaphysical impossibilities which are nevertheless completely intelligible, such that we can make systematic sense of what it would and wouldn't make sense to do in furtherance of various goals were those possibilities realized, then we should think such possibilities are well suited to modeling the contents of belief.

modeling belief, desire, and their relation to action. In particular, it doesn't tell us how to make sense of degrees of belief and degrees of desire, and how *they* relate to action. The only systematic framework for modeling *these* connections is decision theory. But this framework has its own versions of the problems of coarse grain. And unlike in the case of counterfactuals, there are (to my knowledge) no fine-grained, hyperintensional alternatives to decision theory.

As I conceive of matters—more on this in the next chapter—decision theory can be fruitfully seen as what you get if you start with the possible worlds framework for modeling belief and then add some extra detail. Rather than modeling an agent's state of belief merely with a set of possibilities, we model it with a set of possibilities (conventionally labeled "Ω") *and* an algebra[22] of subsets of that set (conventionally labeled "$\mathcal{F}$") along with a probability measure P defined over that algebra. Instead of modeling an agent's desires with a set of desired possibilities (or a set of desired propositions, themselves conceived as sets of possibilities), we assign a degree of desirability—i.e., utility—to each set of possibilities in the aforementioned algebra.[23] We'll also need a set of acts or options available to an agent. Given all these ingredients, we predict that an agent will perform the available act that maximizes expected utility—i.e., she'll perform the action which is such that the sum of the utilities of the possibilities that could ensue if the agent performs the action, weighted by their probability conditional on the agent's performance of the action, is the greatest. While this sketch skirts over many issues on which different versions of decision theory differ, it's enough to generate the problems of coarse grain.

While there are a number of ways of illustrating this, perhaps the simplest is noting that one of the three axioms of probability theory is the unit measure axiom, which requires that the probability of Ω, the set of possibilities whose subsets are assigned probabilities, is 1. And as with the possible worlds framework, probability theory has no way of distinguishing between propositions that are true in the same possibilities. So any proposition true throughout the space of possibilities Ω is identified with Ω, and assigned probabiliy 1 as well. This looks a lot like the problem of logical omniscience—when probabilities are interpreted as degrees of belief, we can interpret the unit measure axiom as saying that an agent is certain of any proposition true throughout the space of possibilities over which her degrees of belief are defined. But more

[22] By an "algebra" of subsets of Ω I mean a set of subsets of Ω that is closed under finite complementation and intersection. It's standard to also require that $\mathcal{F}$ be a σ-algebra, which involves the stronger requirement of closure under countable complementation and intersection.

[23] Sometimes it's a distinct class of "outcomes" that are assigned utilities, as in Savage (1954). The approach I'm sketching here is along the lines of the one in Jeffrey (1965). For present purposes the differences between these two approaches to the foundations of decision theory won't matter.

fundamentally, the structural feature that probabilities are assigned to subsets of the possibility space—leaving the framework unable to distinguish between the probabilities of propositions true in the same possibilities—looks a lot like the problem of deduction.

While there are versions of decision theory that make use of generalizations of probability theory, rather than probability theory as traditionally axiomatized, to my knowledge they all retain analogs of the unit measure axiom, along with the more basic structural feature that it is sets of possibilities to which probabilities (or generalizations thereof) are assigned.[24] And while there are generalizations of probability theory that do not incorporate the unit measure axiom, to my knowledge none have been incorporated into a systematic decision theory, and moreover they still have the structural feature that it is sets of possibilities that are assigned measures, and so would still give rise to the problems of coarse grain, if incorporated into a decision theory.

This may seem like a cheap trick. OK, given how people typically do decision theory, the degreed representation of belief—probability—has various structural features that give rise to problems of coarse grain. But maybe people have just used those representations because they're mathematically tidy, which lets them prove impressive theorems. As philosophers, rather than mathematicians, why can't we just have a picture where degrees of belief needn't obey any structural, probability-like constraints, while holding that they nevertheless combine with degrees of desire to holistically inform behavior? That is, why can't we have a degreed version of the holistic belief-desire-action connection in which the degreed beliefs are not only not probabilities, but are nothing in the neighborhood?

While this question may be asked rhetorically, it has a flatfooted answer. Expected utility isn't well defined, and can't be fruitfully used to model behavior, once the "probabilities" that it takes as inputs needn't obey any structural constraints. Imagine an agent with two options, **Stay** and **Go**. Conditional on **Stay**, she's certain—i.e., assigns as high a degree of belief to it as she assigns

[24] Three quick examples:

1. One family of generalizations of decision theory is "imprecise" decision theory, where instead of assigning a single number to a subset S of Ω, we assign a range—see Halpern (2003, ch. 2) for a helpful introduction. But the values in these ranges are always values that *some* traditional probability measure could assign to S. So when S is Ω itself, the range assigned to S will be $[1,1]$.
2. Another generalization of decision theory uses Dempster-Schafer belief measures, rather than probability functions. (Again, see Halpern, ch. 2.) These, too, obey an analog of the unit measure axiom—in Dempster-Schafer theory, bel(Ω) is always 1.
3. Yet another generalization of decision theory—Buchak's (2013) risk-weighted expected utility theory—modifies the function from probabilities and utilities to favored actions, rather than modifying the probabilities themselves. But because it still uses standard probability theory, it has the same problems of coarse grain as standard decision theory.

to anything—that there will be trouble. She's also certain, conditional on **Go**, that this trouble would double. Sounds like she'll **Stay**, right?[25] But what if she's also certain, conditional on **Go**, that there will be no trouble at all, while she is not certain, conditional on **Stay**, that there will be no trouble at all? If her degrees of belief were probabilities, this wouldn't be possible—it can't be certain that P, conditional on Q, *and* certain that R, conditional on Q, where R is incompatible with P. But without that constraint, there's nothing to rule out this set of degreed beliefs. And given this set of degreed beliefs, it should strike us, I think, as unclear what action to predict the agent will perform. Definitions of expected utility that are equivalent when the inputs are well-behaved diverge when applied to cases like this, and there seems little to recommend one over the other.[26] Our situation here is not unlike that of trying to predict how Lewis will attempt to get from Nassau Street to the railroad, given his inconsistent body of belief.

Stepping back, the problems of coarse grain can seem like problems for an account of the contents of belief, to be solved by adopting a more fine-grained picture of the objects of belief. But if you accept that it's in the nature of beliefs to interact with desires to inform behavior—if you accept a holistic belief-desire-action connection—then you can't respond to the problems of coarse grain merely by adopting a more fine-grained account of belief. You also need an account about how finely individuated beliefs guide action. And I've tried to show that such an account isn't on the table. Given a simple, non-degreed framework for modeling belief and desire, the problem looks difficult, though one *might* be forgiven for hoping that, with the right fine-grained treatment of counterfactuals, it could be solved. But given a degreed understanding of belief and desire, which is for many purposes superior to the non-degreed one, the problem looks impossible. So I conclude that the problems of coarse grain—as they arise for the possible worlds framework for modeling belief, as well as its decision theoretic elaborations—are everyone's problems, rather than a cudgel with which some particular modeling framework can be beaten. In the next section I'll return to the fragmentationist solution to those problems, situating it within the modest approach to modeling more generally.

[25] I've never understood the song. The decision seems simple.

[26] This is one of the lessons I draw from Pettigrew (2021).

3.3.2 Fragmentation as Modest Modeling

Earlier I described the fragmentationist response on behalf of the possible worlds theorist as holding that we model an agent's beliefs not with one set of possibilities, but with multiple sets. In this section, I'll refine that explanation. But first, I want to situate my discussion in the context of some more general ideas in the philosophy of mind. So far I've been describing the possible worlds framework as one that lets us model the contents of belief and other mental states. That suggests that we should think of our psychology—our beliefs, desires, and other mental states—as a real-world system that a possible worlds theorist studies by constructing models of it. While I do think I invited this interpretation, I'll now say that I think it's potentially misleading. Better, I think, to conceive of both informal folk psychology, the possible worlds framework, and the decision theoretic elaborations of that framework, as all the same sort of thing; they're all frameworks for modeling an otherwise intractably complicated biological (or better: physical) system.

This is, I think, a natural reading of the "interpretivist" tradition in the philosophy of mind, associated primarily with Donald Davidson (1973).[27] As I read that tradition, filtered through the lens of a modeling approach to philosophy, it looks something like this. What it is for a system to be an agent is for a certain kind of modeling framework to be able to construct a reasonably well-fitting model of that system. Which modeling framework? Different people give it different names—Dennett calls it "the intentional stance"—but I'll just say "folk psychology." When we use this framework we build models containing agents with beliefs, desires, intentions, and various other mental states, and we predict that those agents will act in ways that make sense in light of those mental states. Such a model "fits" a system to the extent that aspects of its behavior can be more efficiently predicted by going via folk psychological modeling than via other methods.[28]

I have several aims in linking interpretivism with modeling. First, I want to put pressure on a thought mentioned at the end of §3.1.1. This thought says that the neat models constructed by the possible worlds theorist are of course simplified idealizations, in a way that is intended to contrast with the imagined products of some as-yet-undiscovered, improved philosophical psychology. Perhaps possible worlds models are defensible *as* idealizations, this thought

[27] Though it's a big tent; Quine (1960), Lewis (1974), and Dennett (1987) certainly belong, along with many others. See Heal (1997) for some thoughtful discussion of the general family of views.

[28] Or, as Dennett (1981b, p. 59) puts it, "What it is to be a true believer is to be an intentional system, a system whose behavior is reliably and voluminously predictable via the intentional strategy."

goes, but they're very different from what would be delivered by an accurate theory of belief for non-ideal agents like us. If you take interpretivism in the philosophy of mind seriously, and you come to see interpretivism through a modeling lens, then this should look like a mistake. You should think that in describing systems in agential, psychological terms, we've *already* entered the realm of idealization and modeling; a non-idealized theory of belief should sound like an oxymoron. Of course, we can look for modeling frameworks that improve on our current ones, but there are no a priori guarantees about what can be achieved while still using categories at roughly the level of granularity as those of folk psychology.

Second, I hope that the conceptualization of interpretivism I've just offered may make it more palatable. In particular, to those who see interpretivism as objectionably instrumentalist or antirealist—those who say it's a view on which we don't *really* have minds, but it's just useful to talk as if we do[29]—I hope this version of interpretivism will make clear that interpretivism needn't single out *psychological* discourse for such a treatment. Rather, if you think, as I'm inclined to, that a modest modeling approach is called for even in much of the natural sciences, then the idea that the intepretivist is objectionably antirealist may seem to have less bite.[30] Perhaps only atoms and the void are *really* real, and minds are neither. But if that's the only sense in which the interpretivist is an antirealist—if she puts minds in the same "not really real" category as genes, organisms, and species, along with everything else that is neither atoms nor void—is that really so bad?

Lastly, my aim in linking interpretivism and modeling is to set up the main argument of this section; that a modest modeling stance, when the models constructed are psychological, ends up predicting exactly the sort of fragmentation necessary to escape the problems of coarse grain. It's to that argument that I now turn.

Let's provisionally view all mental state attribution—whether the informal, casual sort involved in everyday folk psychological talk, or the more mathematically regimented sort done by Bayesian decision theorists—as a kind of modeling. What would it be to take a *modest* approach to such modeling? It would be to expect that no model of someone in mentalistic terms—no picture of them as a believer, desirer, intender, possesor of a credence function and

[29] For example, Churchland (1983).

[30] For instance, the various "pluralisms" that are popular in the philosophy of biology—see, e.g., Kellert et al. (2006)—all seem to me very much along the same lines as modest modeling. It's often useful to divide up biological reality into different species, but just *how* we make that division will depend on which aspects of biological reality we're hoping to illuminate, and for some purposes—e.g., in classifying bacteria—it may not be useful to talk about species at all (Franklin-Hall, 2007).

utility function, etc.—will fit just right. Every model we can construct will be less than fully informative about someone's behavior and/or make some inaccurate predictions. Moreover, different models will work more or less well for different purposes.

There is a wide variety of examples that have been much discussed in the recent literature on the philosophy of mind that, as I interpret them, illustrate the above phenomenon. Schwitzgebel's (2001) cases of "in-between believing," Tamar Gendler's (2008a, 2008b) cases of "alief" and Stalnaker's (1991) "shrewd but inarticulate chess player," who can play a mean game of chess but can't explain why she makes the moves she does, all fit the bill.[31] They're all cases where no single model of an agent in familiar folk psychological terms—nor any single model in regimented decision-theoretic terms—seems to make sense of all of their behavior. And it's not just that there are aspects of their behavior that we wouldn't have hoped to capture with psychological models—of course we can't rationalize fainting, or vomiting—rather, even when restricting our attention to the sort of phenomena that psychological modeling seems apt to capture, no one model will do. Perhaps somebody talks the talk of an egalitarian, but walks the walk of a sexist.[32] Neither a psychological model in which they have sexist attitudes, nor a model in which they lack them, will capture everything we want. Or they explicitly decide to get on an airplane, but beg to be let off once they're seated. Again, neither a model in which they believe air travel is safe, nor one in which they don't, will fit all the phenomena.[33]

If we are *im*modest modelers, we'll expect that this is merely the sign of the immaturity of the folk psychological modeling framework and its more mathematically regimented descendants. We'll hope that with a better psychological modeling framework, we'd be able to build a single model that would systematically capture both the walk *and* the talk of the implicit sexist, and another one that would capture both the clever play *and* the clumsy analysis of Stalnaker's chess player. Maybe. My suspicion is that as long as we stick to models built using broadly folk psychological categories, there will be serious limits on just how much systematicity we can achieve. And while we could perhaps do better in *some* respects by modeling people in neuroscientific rather than psychological terms,[34] it's hard to imagine such a practice replacing

[31] See also Elga and Rayo (2021) and Rayo (2013a, ch. 4).
[32] See, e.g., Saul (2013) for discussion.
[33] I discuss use this example in Greco (2014c), and offer a fragmentationist treatment of it. Both this example and the previous one will recur in Chapter 6.
[34] As emphasized by Churchland (1986).

talk of belief and desire; in the absence of portable brain scanners, we'll still be left to interpret each other in terms of categories that can be distinguished and deployed without fine-grained neurological information.[35]

So suppose our stance towards psychological modeling is modest; we accept that if we are left to interpret each other using categories that can be applied without fine-grained neurological information, then the models of each other that we build using those categories are doomed to be limited in scope and fit, as in the examples above. This, I suggest, is how we should understand fragmentation, as it figures in the response to the problems of coarse grain.

Let's return to Lewis and his beliefs about the streets in Princeton. Using the concepts and categories I've been developing, here's how we can interpret the suggestion that Lewis' "system of beliefs was broken into multiple (overlapping) fragments," and that "different fragments came into action in different situations." No one folk psychological or decision theoretic model could neatly capture all of Lewis' behavior. Depending on the situation, one or another model will fit best. Lewis doesn't tell us about which situations elicit which "fragments" of his body of belief, but we can flesh out the story in a variety of ways. Perhaps it depends on which street he thinks of first; if he thinks of Nassau Street first, he thinks it runs east-west, and then infers that the roalroad must too because they are parallel. And if he thinks of the railroad first, he thinks it runs north-south, and then thinks Nassau Street must too, because they are parallel. Or perhaps it depends not on which street he thinks of first, but on how he's traversing the terrain—on foot he imagines the town map one way, while in a vehicle, another. Or maybe there's no tidy, systematic way of specifying when Lewis is neatly modeled as having one body of beliefs, and when another.[36]

In the language of modeling, for an agent to have a "fragmented" body of belief or desire is for folk psychological modeling of that agent to go best when we don't insist on sticking to a single model, but instead use different

[35] As Dennett (1991) points out.

[36] When there's a sufficiently systematic way of saying which body of information (modeled as a set of possible worlds) guides the agent's behavior, we might want to say that instead of using various different possible worlds models, we use a single, complex model that involves something like a function from situations or "elicitation conditions" to bodies of information. This is the approach taken by Elga and Rayo (2021), and is in the spirit, I think, of the picture in Parikh (2008), where the problem of logical omniscience is addressed in a broadly fragmentationist spirit by thinking of bodies of belief as relative to action plans—i.e., I might believe P relative to one action plan, but fail to believe it relative to another. I haven't put things this way because I want to emphasize the continuity between cases where there is a systematic story about which possible worlds model to use to capture someone's behavior (whether in terms of plan-relativity, or elicitation-condition-relativity, or something else) and cases where things are messier.

models in different situations.[37] We can also put the idea slightly differently, in interpretivist terms. The interpretivist says that the mental states an agent in fact has are the ones that a certain kind of interpreter would interpret her as having. Traditionally, the interpreter is understood as facing various constraints—she wants the beliefs and desires she attributes to fit the agent's behavior, to evolve over time in rational ways, to track natural distinctions,[38] and perhaps more. One way of interpreting the fragmentationist idea is that we simply relax the constraint the interpreter must assign a *single* body of mental states that best meets all these constraints. Fewer is certainly preferable, and perhaps the behavior of an ideally rational agent could be captured with a single body of mental states; her behavior would be sufficiently coherent that the interpreter's job would be easy. But if, as is likely typical for creatures like us, each single body of mental states does very poorly by the lights of *some* constraints, while attributing multiple sets—each of which fits different aspects of the agent's behavior—does much better, then the psychological truth about the agent is that she has multiple sets of mental states, i.e., is mentally fragmented.

We've already seen that the problems of coarse grain for the possible worlds framework—and, by extension, its decision theoretic elaborations—can be addressed if we don't insist on modeling agents with single models built from that framework, but instead can pick and choose, capturing some aspects of their behavior with one model, and other aspects with another. What I've tried to show here is that this approach to psychological modeling is just a special case of what, in the previous chapter, I suggested is typical when building models using highly non-fundamental categories.

My aim in this chapter has been to show how the possible worlds framework for thinking about information and the idea of modest modeling form a mutually reinforcing package. The possible worlds framework for modeling content has been undeniably fruitful across an extremely wide range of applications, but it faces the perennial problems of coarse grain. Should we abandon it on that basis? We should not. A good Kuhnian will only abandon a paradigm for a superior rival, and as yet no alternative to the possible worlds framework shares its virtue of being able to model the way in which information and value together rationalize action. Moreover, the problems of coarse grain only look so problematic if you start out with some ambitious assumptions

[37] To be clear, a case where we have a single model in which an agent's belief change over time, e.g., via conditionalization, isn't what I'm talking about.
[38] Lewis (1983b).

about what psychological models should do. If you take a modest approach to psychological modeling, then the familiar "fragmentationist" reply to these problems ceases to look desperate and unmotivated, and instead just looks like what we should have expected all along.

3.4 Coda: Model Contextualism

In this section I'll explore the connection between the fragmentationist, modest modeling approach to mental states, and questions about mental state *attribution*. For instance, if we think of belief along the lines I've been suggesting, what if any commitments do we incur concerning which sentences of the form "S believes that P" are true?

It's initially tempting to think we're pushed towards some kind of relativism or contextualism about mental state attribution. If a modest modeling picture is right, then perhaps claims of the form "S believes that P" aren't true *simpliciter*, but only true relative to a set of elicitation conditions,[39] or an interpretive framework, or a model.[40] Or perhaps they are true *simpliciter*, but only because whenever they're uttered context "fills in" a value for some hidden parameter along the lines of the previous, in the same way that it's typically held that context fills in location and time parameters that allow for utterances of "it's raining" to be true when and only when uttered in locations where it's raining at the time of the utterance.[41]

This is too quick. First, as I'll explain below, it's possible to combine a fragmentationist view of belief with a non-relativist, non-contextualist theory of belief attribution, as Lewis arguably does. But more fundamentally, I think it's a mistake to think that a fragmentationist picture of mental states should push us towards a distinctive view about how to do the semantics of belief attribution. As I've already discussed, I see fragmentationist views of mental states as a special case of modest modeling, which is an attractive approach to take not only to model-building in philosophy, but also in economics, biology, and many other areas. But it would be surprising if a methodological debate among economists should have consequences for how linguists ought to model the semantics of words like "price" or "growth." Likewise, if a semanticist becomes convinced by philosophers of biology that she should adopt a pluralist

[39] E.g., Elga and Rayo (2021).
[40] This is inspired by MacFarlane (2011) style relativism.
[41] See, e.g., Kaplan (1977/1989).

view about gene concepts—more on this later—that doesn't suggest that, qua semanticist, she should treat "gene" differently from any other count noun. Ultimately, I think the same goes for fragmentationism about mental states, and sentences of the form "S believes that P." More generally, I think taking a modest modeling approach towards some phenomenon of philosophical interest—belief, knowledge, etc.—can often let us mimic some of the advantages of contextualism or relativism, but without the need to posit that discourse about that phenomenon is semantically unusual or distinctive. I'll call the resulting position "model contextualism," but as will emerge, whether the position really deserves to be called a version of contextualism is far from clear. But I'm getting ahead of myself. Before I explain and defend the remarks I just made, it will help to see how one can be a fragmentationist without going in for anything like contextualism about belief attribution.

Lewis himself is naturally interpreted along these lines. I'm inclined to read him as accepting something like the following view:

∃Fragment: "S believes that P" is true just in case P is true according to *some* fragment of S's system of beliefs.

∃Fragment is a non-contextualist, non-relativist view about belief attribution, as I intend it to be interpreted. On this view, while individual fragments of an agent's system of belief will, in one situation or another, guide an agent's behavior, it's always the whole system that determines the truth of belief attributions. So to determine whether "S believes that P" is true, we don't have to first figure out which fragment we're talking about—it's true just in case there exists a fragment of S's beliefs according to which P is true.[42] I'm not too concerned with whether this really is the right interpretation of Lewis; for my purposes, it's useful enough as an illustrative example of the independence of the fragmentationist picture on the one hand from specific commitments concerning the semantics of belief attributions on the other.[43]

Given that views like **∃Fragment** are available, why might a fragmentationist hesitate to embrace it? Let's return to the example of the implicit sexist. In many contexts—e.g., when we're discussing his performance in a public discussion—we can felicitously explain his behavior by attributing to him the

[42] The extension to desires, credences, and utilities is straightforward.

[43] For what it's worth, I think this is clearly the right interpretation of a perhaps semi-technical locution Lewis introduces: "true according to S's beliefs." Whether Lewis thought "S believes that P" should get the same treatment as "P is true according to S's beliefs" is a question on which I see little textual evidence one way or the other, and with which I'm not myself particularly concerned.

absence of sexist beliefs. E.g., if someone asks why he didn't say "yes" when asked whether men are generally better at abstract reasoning than women, we might truthfully respond: "because he doesn't believe that." And this is so *even if* in other contexts—e.g., in explaining why, in advanced logic seminars, he tends not to call on women, and tends to be dismissive and skeptical of their contributions—we can explain his behavior by attributing to him that very belief.

More generally, it's plausible that the information we aim to convey in attributing the presence or absence of a belief to someone is typically more local and specific than information whose truth or falsity is sensitive to someone's total fragmented belief state. To take another example, imagine you're producing a television show in which commentators will discuss sporting events. You plan to hire athletes as commentators. With an eye towards producing engaging discussion with a lively clash of perspectives, you want to hire commentators with a mix of views about proper strategy—perhaps you'll want some who favor a "sabermetric," big-data approach, and others who favor more traditional, intuition-based approaches. In your discussions about whom to hire, you'll attribute various mental states to potential commentators. But crucially, you'll only be interested in mental states that could be manifested in discussion. Imagine one of your potential hosts talks the big data talk, but in his gametime, split-second decisionmaking, his behavior is naturally interpretable as guided by traditional beliefs about proper play. Perhaps he *says* that statistical analysis reveals that baseball players attempt to steal bases far too often, and that the risk typically isn't worth the reward. Nevertheless, when he's on the field, he's just as likely as the average player to attempt to steal a base. Given the views defended in this chapter, he counts as fragmented. But in the context of a discussion about his suitability as a commentator, the beliefs about proper play that are manifested only in his gametime decisionmaking and not at all in his commentary are simply irrelevant, so it's natural to interpret belief attributions in the context of such a discussion as being somehow implicitly fragment-relative; when we say that he doesn't think players should steal bases, we're not taking on any commitments about how he would behave on the field—only about what he would say in a discussion. And of course things would be entirely reversed if we were evaluating him as a player, rather than a commentator. So does this mean we should opt for some kind of semantic contextualism or relativism about belief attribution? For reasons alluded to earlier, I don't think so, or if it does, it shouldn't push us towards a form of contextualism or relativism that distinctively concerns belief attributions.

To explain why, I want to take a detour through philosophy of biology. As Rheinberger and Muller-Wille (2015) write, "More than a hundred years of genetic research have rather resulted in the proliferation of a variety of gene concepts, which sometimes complement, sometimes contradict each other." While each of these categories could themselves be broken down further, one taxonomy involves distinguishing units of function, units of mutation, and units of recombination.[44] These aren't the same—e.g., while a gene understood as a unit of function must code for a protein, a gene understood as a unit of mutation need not. For different theoretical purposes—e.g., theorizing about rates of mutation and accumulation of "junk" DNA, versus theorizing about the genetic bases of superficial traits—biologists will identify and distinguish genes differently. So syntactically identical sentences involving "gene" can be used to convey different pieces of information in different theoretical contexts. We might call this a version of contextualism, and some philosophers of biology do. At the end of a paper in which he's argued for a kind of context-dependence even within the concept of the gene as a unit of function—he focuses on cases of "molecular pleiotropy, in which distinct molecules are derived from a single putative gene"—Richard Burian (2004) writes the following:

> Covertly or overtly, but unavoidably, the principles on which genes are identified and individuated are context dependent.... Given this, the muddiness of gene concepts, frustrating to many philosophers, but celebrated by a few, is inevitable. If this strong version of context dependence is correct, the continuing evolution of gene terminology is not going to stabilize at some new orthodoxy based on a strictly intrinsic characterization of genes.
>
> (p. 77)

I find it helpful to view this kind of "contextualism" about genes as a species of modest modeling. It's very natural to view Mendelian genetics as a kind of modeling framework. Want to understand and predict the distribution of stem lengths or seed shapes across multiple generations of peas? Mendelian genetics provides a framework for doing so, and that framework involves modeling individuals as having genes that correspond to observable, phenotypic traits. Part of the reason it's so natural to think of this as a modeling framework as opposed to something more lofty is that, as genetics has progressed, we can now see a host of ways in which Mendel's approach involved what in retrospect

[44] This is how Burian (1985) summarizes an older classification due to Seymour Benzer.

look like idealizations. Each gene corresponds to a single trait, each trait is controlled by a single gene, genes act independently, etc. But, at least according to Burian, and contra the "traditional" approach to modeling described and rejected by Batterman (2009), the progress of genetics has not converged on a single, more sophisticated and detailed successor to Mendel's initial efforts. Rather, if you want to use genes to do explanatory work, there are a host of different modeling frameworks you might employ, each of which will be fruitful for explaining different aspects of biological reality. Genomes don't come pre-sorted into genes and, depending on what you want to illuminate, you'll do the sorting differently.

We can say all this, I think, without yet taking on commitments about how, from the standpoint of linguistic semantics or philosophy of language, "gene" talk should be treated. An orthodox contextualist approach, as I understand it, would run as follows. "Gene" is an indexical, like "I," "here," and "now." While the content of those words (in context) depends on the speaker, location, and time of the context, the content of "gene" in context depends on, perhaps, whichever specific gene concept is salient in the conversation when the word is uttered.

This strikes me as an unattractive route. It's often presupposed that, insofar as a theorist is going to posit that a term is an indexical—that it belongs in a class with "I," "here," and "now"—she bears the burden of showing that speakers are implicitly aware of the indexicality of the term. This is where the famous "semantic blindness" objection to epistemic contextualism comes from: epistemic contextualists seem to posit contextual parameters for "knows" which speakers are ignorant of, and that's thought to count against their view.[45] Defenders of epistemic contextualism often take up the challenge, and argue that ordinary speakers in fact are implicitly aware of the context sensitivity of "knows."[46] If we accept the presupposition that speakers should be (at least implicitly) aware of indexicality in their language, then "gene" seems like a very poor prospect for an indexical. Ordinary speakers use the term, but have no idea about any of the complexities philosophers of biology discuss. And it seems to me overly optimistic to assume that ordinary speakers semantically defer to geneticists who themselves are (at least implicitly) aware of the context-sensitivity of "gene."[47] The proliferation of gene concepts has happened gradually over the course of a century, and geneticists needn't be and typically aren't also semantic historians.

[45] Hawthorne (2004). [46] See, e.g., DeRose (2005). [47] See, e.g., Burge (1991).

So what *should* we say about the semantics of "gene," once we are modest modelers? One response would be to say that this is all a matter for pragmatics. The larger the gap we're willing to tolerate between what speakers typically use sentences to communicate, and what those sentences mean, the less we'll be inclined to think that the fact that sentences involving "gene" will be used to convey different information when uttered in the context of different theoretical projects has any bearing on the semantics of those sentences. Maybe the semantic contribution of "gene" to sentences it occurs in is invariant, and pragmatics will explain how speakers can use those sentences to mean different things in the different contexts.[48]

While I have some sympathy for this position, and I suspect the applications of model contextualism I'll make throughout this book would be largely unchanged if the "contextualism" in question were an entirely pragmatic phenomenon, I want to suggest a different route. Semantics itself is a model-building discipline *par excellence*,[49] so rather than asking which of some existing menu of semantic options for some expression is suggested by taking a modest-modeling approach to the phenomena that expression is used to talk about—e.g., which menu of semantic treatments for "gene" is suggested by taking a modest modeling approach to genetics, or which menu of semantic treatments for "believes" is suggested by taking a modest modeling approach to mental states—it may be more fruitful to ask what happens when we take a modest modeling approach to semantic theorizing itself.[50]

Orthodox contextualists see a sentence used to convey different information in different contexts, with correspondingly varying intuitive truth-conditions, and try to capture that variability in a semantic model of the behavior of (some of) the words in that sentence. In the model, sentences are uttered in contexts, and which propositions those sentences express systematically depends on certain specifiable features of those contexts.[51] We can imagine an alternative approach on which, rather than building context-sensitivity into our semantic models, we let our choice of a specific semantic model be guided by context. So rather than modeling "gene" as picking out a function from contexts to extensions—thereby building context-sensitivity into our semantic model—instead, depending on our own context, we'll model "gene" as picking out

[48] This is how I understand Kent Bach's view (2008). He argues that in many cases where philosophers argue that the meaning of a sentence is context-dependent, in fact what's going on is that the meaning of the sentence is invariant but quite thin, and it's only in context, and after pragmatic enrichment, that an utterance of that sentence manages to express a truth-evaluable proposition.

[49] See Yalcin (2018).

[50] Doing so is very much in the spirit of Agustín Rayo's "A Plea for Semantic Localism" (2013b).

[51] See, e.g., Lewis (1980).

different extensions.[52] For instance, if we're using semantic models to shed light on a linguistic exchange in which rates of mutation are being discussed, we'll model "gene" as having a different extension than if we're using semantic models to shed light on a linguistic exchange in which the genetic bases of phenotypic traits are being discussed.

What's the difference between building context-dependence into our semantic models, as the orthodox contextualist does, and instead just using different semantic models in different contexts, as the modest modeler does?[53] As I understand it, there are several. First, the modest modeler isn't committed to their being specifiable features of contexts on which her choice of models depends. She needn't claim to have a systematic story about which precise semantic model she'll use when. Rather, like the economist who's comfortable relying on rough and ready heuristics to decide which model of economic growth to use to predict the trajectory of this or that economy, the modest (semantic) modeler can say that she knows she'll use different models in different contexts, without being able to describe a function from specifiable features of the context to her choice of model. Second, by not treating context dependence as a feature of the meaning of a word, worries about "semantic blindness" quite clearly don't apply. A modest (semantic) modeling approach to "gene" does *not* model "gene" as having a context-dependent meaning that speakers are mysteriously ignorant of. Rather, in any given context, the approach will model "gene" as having a fixed, context-independent extension.

Of course, this invites its own, distinct objections. Isn't there some *explanation* of why "gene" can be used to refer to one sort of entity in one context, and another in another? And doesn't that explanation somehow depend on the meaning of "gene," in a way we should hope would be captured by semantic modeling? The orthodox contextualist can easily answer analogous questions about "I"—the reason "I" is used to refer to me when I say it, and to you when you say it, is that whenever anybody says it, it refers to the speaker. Shouldn't there be some closely analogous story we can tell about why "gene" sometimes refers to these stretches of DNA, sometimes to those?

[52] I'm presupposing that count nouns like "gene" have as their semantic values extensions. For those who favor a "thinner" approach to semantic values, on which the semantic value of a sentence typically doesn't determine truth conditions, this will likely seem unattractive. Here again I think modest modeling *may* allow for a kind of reconciliation, though it would be beyond the scope of my project to provide it. It may be that for some purposes, it's illuminating to model meanings as "thin," while for others, modeling them as rich enough to determine truth conditions is fruitful.

[53] To be clear, I don't mean to be treating modest modeling as an across-the-board alternative to contextualism. My own inclinations are to adopt an orthodox contextualist treatment of uncontroversial indexicals like "I," "here," and "now", but a model contextualist treatment of "gene" and "believes."

Perhaps, but perhaps not. It may be that the kind of systematic explanation we'd like of why "gene" can't be happily modeled with a single, fixed extension, can't be given within a modeling framework in which words are assigned extensions. To return to the cases closer to my heart, it's tempting to think there must be some explanation of why Lewis sometimes behaves as if the streets of Princeton are arranged one way, and other times another. While I'm no believer in the principle of sufficient reason, Lewis' navigational quirkiness doesn't seem like a plausible candidate for a brute, inexplicable fact. And yet it might be brute and inexplicable *from the perspective of folk psychology*. That is, it may be that in order to explain why Lewis is sometimes happily modeled as believing one body of information, and other times another, you need to move away from a folk psychological framework in which we model agents as having beliefs. Maybe we need to go down to a neuroscientific level, and use a different set of categories.[54]

My suspicion is that (a) this is true, and (b) an analogous claim holds concerning semantic modeling of "gene," and other expressions from theoretical projects for which modest modeling stances are attractive. That is, there *is* some explanation of why "gene" is sometimes best modeled with one extension, sometimes another, but it's not one that can be given within the very same modeling framework in which count nouns are modeled as having extensions as their semantic values.

What I'm calling a "model contextualist" approach to an expression is meant to have the advantages of theft over honest toil. With the orthodox contextualist—in particular, someone who thinks fragmentation about the mental supports orthodox contextualism about mental state ascriptions—the model contextualist will say "yes that's true" when somebody ascribes a sexist belief to the implicit sexist while discussing his classroom management practices, and will also say "yes that's true" when somebody ascribes the absence of that sexist belief in the course of a discussion about his performance in a debate. These are the advantages. But as for theft, she'll deny that she thereby incurs a debt to provide a distinctive semantic model of "belief" that would vindicate her linguistic practice. Perhaps such a debt *could* be paid—model contextualism is consistent with orthodox contextualism—but the model contextualist denies that she owes it in the first place. If asked to defend herself, she'll say that talk of "belief" is part of a folk psychological framework for which modesty in model selection is appropriate, and thus she gives up no more hostages to semantic fortune than do biologists, economists,

[54] If you're really pessimistic, you might think we have to go all the way down to physics.

and other modelers who use different, incompatible models at different times to illuminate different aspects of their respective domains.

Orthodox applications of contextualism in philosophy take sets of intuitively plausible but apparently inconsistent assertions, each made in different contexts, and claim that the right semantic model of the meaning of some of the words used to make those assertions will reveal that the assertions ultimately don't express inconsistent propositions. So we can happily regard each of the intuitively acceptable assertions as expressing truths. My strategy at various points in this book will be similar—I'll take sets of intuitively acceptable but apparently inconsistent assertions, each made in different contexts, and argue that we needn't reject any of the attractive assertions. But I won't do that by providing unifying semantic models on which all of the assertions are true.[55] Rather, what I'm calling "model contextualism" amounts to a kind of companions in guilt response. Just as economists, biologists, and other modelers can seem to be saying inconsistent things when they use different, incompatible models in different contexts—and perhaps they are—this sort of inconsistency isn't something responsible inquirers should strive to eliminate. Moreover, this judgment about methodology requires no semantic vindication—just as number theorists can get on with their business without worrying about the ontological status of numbers, modelers in the special sciences should be able to do their research without waiting on semantic models of "gene," "species," or "money" that would vindicate their talk about these matters.

Of course there are limits to the applicability of this kind of maneuver—we don't want an "anything goes" methodology in which one can always defend oneself against the charge of inconsistency by saying: "I was using a different model back when I said that." But as I see it, the question of when this response is legitimate and when it instead amounts to special pleading essentially depends on which theoretical alternatives are available. I can only legitimately make the model contextualist move when there isn't a competitor framework that can offer a unified explanation of phenomena I need multiple models to accommodate. In my judgment, that's true of the various applications of model contextualism I'll make throughout this book. But readers will have to decide for themselves whether I'm right about that.

[55] I'm happy to offer a collection of flatfooted, contextually invariant semantic models, each of which would vindicate *some* of the assertions, but this just pushes the problem of inconsistency back: those semantic models will be incompatible with one another.

4
Certainty and Undercutting

A familiar potted history of modern epistemology runs as follows. Descartes gave us the project of building our edifice of belief upon a foundation of indubitable certainties. But the Cartesian project was an irredeemable failure; there are either no indubitable certainties or too few to serve as an adequate foundation for the rest of what we reasonably believe. The central project of post-Cartesian epistemology, then, has been to show how we can think about knowledge, belief, justification, evidence, and inquiry, without appeal to any certainties. Some have thought the problem was with Descartes' view about the structure of knowledge. Rather than the metaphor of a building resting on a foundation, a better metaphor for the structure of knowledge would be that of a web,[1] where no individual strand is particularly important and the system is sound because of how the individual elements reinforce each other. Others were comfortable with the idea of belief having a hierarchical structure, with foundational elements supporting non-foundational elements, but emphasized the fallibility or provisionality—the uncertainty—of even the foundational elements.[2] But all post-Cartesian epistemologists can agree that we shouldn't call on certainty to do any heavy philosophical lifting.

In light of this history, it's an awkward fact that fruitful formal models of belief seem to make liberal use of certainty. In Bayesian epistemology, learning is typically modeled as a process in which an agent becomes certain of some new fact. In linguistics and philosophy of language, it's standard to model assertions as having the effect that, going forward, propositions asserted are treated as certain by their audience.[3] In game theory, while uncertainty is important, so too is a backdrop of certainty; agents are typically modeled as certain about various features of their strategic situations; e.g., what possible strategies they and the other players might choose, and what outcomes would result in each state depending on each player's choice of strategy.

[1] E.g., as in Quine and Ullian (1978).

[2] For example, Gilbert Harman (2001) defends a view on which *all* beliefs count as foundational. Phenomenal conservatives defend the only slightly less permissive view that all beliefs that *seem* true to their subjects are foundational for those subjects.

[3] As in the literature inspired by Stalnaker (1978).

Idealization in Epistemology: A Modest Modeling Approach. Daniel Greco, Oxford University Press. © Daniel Greco 2023.
DOI: 10.1093/oso/9780198860556.003.0005

The received view of this tension—to the extent that there is one—is that treating beliefs as certain is a fruitful but false idealization.[4] Strictly speaking, learning involves the acquisition of near-certainties, assertions are accommodated by people becoming nearly certain of their contents, and people playing games are confident but not certain of the structure of the games they're playing. But models that explicitly incorporated all this uncertainty would be unwieldy, so it's often fruitful to round confidences up to 1 or down to 0.

I think this received view is half-right. My take on certainty in this chapter will parallel my take on logical omniscience in the previous one. While I agree with the received view that models incorporating certainty (logical omniscience) are idealized, I deny that there must exist non-idealized models according to which all those certainties are uncertain (or, according to which we have beliefs and knowledge that are not closed under logical entailment). If we are ambitious epistemological modelers, we'll think that out there in Plato's heaven there are the true models that capture exactly what attitudes we do (descriptive) and should (normative) have towards all propositions, and those models will feature few or no certainties. But if we're modest modelers, we can be happy using models in some contexts in which it's certain that P, and using models in other contexts in which it's not certain that P, without holding out hope that there's a grand model that all these little models are approximating. Perhaps the only "strict" truths in the neighborhood would have to be stated in the language of neuroscience;[5] once you're modeling someone as a believer, there's no single model that works best for each context, and the little local models that work well in little local contexts may each involve a hefty dose of certainty.

That's the birds-eye view. To make the case, I'll start in §4.1 by giving some examples of where certainty shows up in formal models of belief, focusing on Bayesian updating. In §4.2 I'll review the familiar case for thinking these appearances of certainty should be understood as idealizations, along with a couple of attempts to show how we might eschew those idealizations. In §4.3 I'll then suggest that the difficulties that arose in §4.2 can be addressed by being modest about model selection, rather than by looking for frameworks that let us build models in which nothing is certain. Lastly in §4.4 I'll consider an objection to the effect that the view described in §4.3 sits uneasily with

[4] For example, Moss (2019) holds that when we describe people as believing things, we're ascribing certainty to them, and these ascriptions are false. Loosely speaking, we are certain in various claims, but strictly speaking we are certain in almost nothing.

[5] Or physics.

the practice of using normative models in service of first-person deliberation about what to think or do.

4.1 Certainty in Bayesian Models

Certainty is deeply baked into the Bayesian framework. Recall that in this framework, we represent an agent's opinions with a set of possibilities Ω, an algebra $\mathcal{F}$ of subsets of that set (these represent the propositions or events about which the agent has opinions), and a probability function P that assigns numbers to the sets in $\mathcal{F}$; the higher the number assigned by P to a member of $\mathcal{F}$, the more confident the agent is in the proposition represented by that subset.

We saw in the previous chapter that a structural feature of this framework leads to some inevitable certainty; the agent must be certain of the proposition represented by Ω, the space of possibilities over which she distributes her opinions. Maybe this isn't so bad; if the space is big enough—if it includes possibilities where gravity stops working tomorrow, and possibilities where the agent is being deceived by an evil demon, and possibilities where... (you get the idea)—then perhaps to be certain of Ω isn't to rule out much. But it's worth dwelling on the fact that in this framework, certainties aren't an optional add-on. If you want an agent to be *un*certain of whether P, you need to make sure you include both P-possibilities *and* not-P-possibilities in Ω. Representing uncertainty takes work. Usually we won't make sure to include lots of recherché possibilities in our models, and when we do less such work we thereby represent more as certain. For instance, if we want to represent an agent's attitudes about some game of chance, we'll typically model her uncertainty using a space Ω that includes the possible outcomes of the game, and that's it. If an agent is playing a game in which a die will be rolled, we might represent her opinions using a model in which Ω is $\{1, 2, 3, 4, 5, 6\}$, the space of possible outcomes of the roll of the die. But to use this space is to model the agent as certain that the die will be rolled. Later in this chapter I'll address the view that this should be thought of as an eliminable idealization—we *could* add in possibilities where the die isn't rolled, it's just often not worth the effort—but for now I just want to build up a little catalog of some ways in which certainties pop up in models of belief.

Let's set aside an agent's certainty in Ω, and move to the question of learning. The standard way of representing learning in the Bayesian framework is for there to be some subset of Ω that the agent learns to be true with certainty.

E.g., our dice-playing agent might learn that the die landed even. The rule of conditionalization tells her how to move from a state of opinion in which the agent thinks any of {1, 2, 3, 4, 5, 6} might be true to one in which she's certain that one of {2, 4, 6} is true. The basic idea is that the ratios between the probabilities of the remaining live possibilities should remain unchanged, but the total probability of the remaining possibilities should increase so that their sum is 1. So if our agent initially regarded each of the possibilities in {1, 2, 3, 4, 5, 6} as equally likely, assigning a probability of $\frac{1}{6}$ to each, then when she learns the die landed even she should continue to regard each of the possibilities in {2, 4, 6} as equally likely—i.e., she should hold fixed the ratios of the probabilities of 2, 4, and 6—which will now require her to assign $\frac{1}{3}$ to each of those possibilities.

Here too, certainty is deeply baked into the framework. In representing the agent as moving from a state of opinions distributed over {1, 2, 3, 4, 5, 6} to one distributed over {2, 4, 6}, we've thereby represented her as becoming newly certain that the die landed even. If you want to represent learning something, but not with certainty, you need a different way to model learning.[6] While this may seem like a flatfooted point, it's worth belaboring, as in the popular consciousness the Bayesian framework for thinking about learning is often associated with the idea that nothing is certain. For instance, Sean Carroll, writing in the internet magazine *Edge*, proposed "Bayes' theorem" as his answer to their 2017 question: "What Scientific Term or Concept Ought to be More Widely Known?" After an accurate exposition of the theorem, he offers the following remark:

> The other big idea is that your degree of belief in an idea should never go all the way to either zero or one. It's never absolutely impossible to gather a certain bit of data, no matter what the truth is—even the most rigorous scientific experiment is prone to errors, and most of our daily data-collecting is far from rigorous. That's why science never "proves" anything; we just increase our credences in certain ideas until they are almost (but never exactly) 100%.[7]

But as I've just explained, in the orthodox Bayesian framework, in representing a change in an agent's confidence as prompted by new data, you thereby

[6] I'll discuss the generalization proposed by Jeffrey (1965) soon.
[7] https://www.edge.org/response-detail/27098.

represent the data itself as certain. Of course, there are powerful reasons to think of this as some kind of idealization—I'll come to them shortly—but for now I just want to emphasize that, far from the Bayesian framework explaining or illuminating the idea that nothing is certain, it seems to be in tension with that idea. While I ultimately think we can have our cake and eat it too—we can regard the Bayesian framework as a good one for modeling learning, while still accepting the spirit of the post-Cartesian thought that nothing is certain—doing so will take some tricky interpretive work!

Before getting to that work, I want to note a parallel between certainty in models of belief updating on the one hand, and in models of conversation on the other. At least since Stalnaker (1978), models of conversation in linguistics and philosophy of language have looked a lot like models of belief updating. In the Stalnakerian framework, the aspect of conversation that comes in for modeling is informational exchange. The context of the conversation is modeled by a set of possibilities—these represent the possibilities treated as live by the parties to the conversation—and it is updated when speakers make assertions. The Stalnakerian model of how assertions update the conversational context is closely analogous to the Bayesian model of how learning updates a belief state; in the Stalnakerian framework, the effect on the conversational context of a speaker asserting that P is essentially the same as—in the Bayesian framework—the effect on an agent's belief state when the agent learns that P.

And so, like Bayesian models of learning, Stalnakerian models of conversation involve some baked-in certainty. When I say "the die landed even," and you accept my assertion, a Stalnakerian model of our conversation would represent us as treating it as certain—for the purposes of the conversation at least—that the die landed even. To be sure, the framework has been generalized to allow for hedged or qualified assertions. I might instead say "the die probably landed even," and we'll want a way to represent that this updates the conversational context *not* by making it certain that the die landed even, but only by making it probable.[8] But the fact that not all assertions are hedged in this way underscores that the most natural way of modeling ordinary, unqualified assertions is to represent them as updating the context in such a way that, going forward, the parties to the conversation treat the content of the assertion as settled, i.e., certain.

[8] See, e.g., Swanson (2016), Moss (2018).

4.2 Certainty as Idealization

Why think of the certainties baked into models of belief, learning, and conversation as mere idealizations? We can distinguish two main reasons. First, there's the fact that for just about any putative certainty P, we can imagine ways we might turn out to be wrong about P, and which we seem unable to conclusively rule out; the Cartesian strategy of dislodging certainty via the use of skeptical scenarios is highly effective. Second—and it's not clear to me how distinct these two reasons really are—we can appeal to considerations having to do with rational action. We've already seen how Bayesian models of rational belief are a proper part of Bayesian models of rational action. And in such models, if it's certain that P, then facts about how well or badly some course of action could turn out in the event that P is false make no difference to the choiceworthiness of the action. For example, suppose for *reductio* that it's certain your home won't be flooded. And suppose we're evaluating the action of buying flood insurance. If it's certain that your home won't be flooded, then how willing you will or should be to buy flood insurance will be entirely insensitive to just how much worse it would be to face a flood without insurance than with it. But that seems wrong; if it would be *much* worse to face a flood without insurance than with it, that fact counts in favor of buying insurance, and if you're rational you'll be sensitive to that difference. Set the numbers right—make the insurance cheap and generous enough, and the prospect of a flood bad enough—and buying insurance will seem like the prudent course of action. But Bayesian decision theorists can't say this if they also say it's certain that your home won't be flooded.

By and large, the response to such examples is to insist that nothing (or very little) is really certain. If you wouldn't bet your life against a penny that P, then you're not certain that P, and since there's little or nothing you'd bet your life against a penny on, there's little or nothing of which you're certain.[9]

So, the thought goes, if we neither are nor should be certain of much of anything, and yet standard models of belief, learning, and conversation

[9] This argument can also be given in a normative key, in terms of what you *should* bet your life against a penny on. See Clarke (2013) for a helpful summary of writers who've made arguments along these lines, along with a contrary take on what the arguments show. I myself made such an argument in Greco (2013). Williamson (2000) is an interesting outlier. He is happy to use the Bayseian framework to model evidential support, but rather than respond to such examples by holding that (strictly speaking) little or nothing is maximally supported by our evidence, he instead holds that there's no straightforward connection between levels of evidential support and rational betting behavior. That is, he's happy to use Bayesian models in epistemology, while remaining skeptical about Bayesian decision theory. See my Greco (2013) for discussion of Williamson's view, though as this chapter should make clear, my own view has changed a great deal since then.

represent agents as certain of a great deal, then those models are thereby idealized. In that case, what can we say about the non-idealized truth that these models are approximating? In this section I'll consider two species of answer to that question, before ultimately suggesting modesty about Bayesian models as a balm for worries about the role that certainties play in them. On one, it's true that learning involves becoming certain of new facts, but the new facts are always restricted to facts in which we *really can* be certain, unlike facts about dice. On the other response, learning involves shifting probabilities without adopting any new certainties.

Which sorts of facts could we rationally learn with certainty? Our options are limited. If you tell me the die landed even, it could always turn out that you were lying. So maybe what I learn with certainty is that you *said* the die landed even. But that won't do either; I could've misheard you. Maybe what I learn with certainty is that you said something that sounded like "the die landed even." But even then, perhaps I was suffering from auditory hallucinations and you didn't say anything at all. Or I could've dreamed the whole thing.

The usual move at this point—probably inspired by Descartes' *cogito*—is to look for some claim about my mental life that seems immune to skeptical threats, and which supports the claim that the die landed even. Perhaps I'm experiencing sense data of a certain character, or it *seems to me* that you said the die landed even, or something along these lines. For instance, Smithies (2019a) offers a recent defense of the idea that ideally rational agents update their beliefs by learning—with certainty—claims about the character of their phenomenal consciousness.[10] Before getting into more principled reasons for pessimism, I'll note that even granting the friendliest understanding of what such claims would have to mean—a concession I'll take back in next paragraph—it's not at all obvious that we really can be certain of how things seem to us, or what we're conscious of. Psychologists and philosophers debate questions about just what the content of conscious experience is, and in such debates it can often seem like we have much better epistemic access to various external world facts than to (some) facts about our own conscious experience. For instance, it's easier to know what subjects report in cognitive scientific experiments than it is to know just how rich and detailed my conscious experience is of objects in the periphery of my visual field.[11]

[10] I respond to Smithies in "Accessibilism without Consciousness" (Forthcoming).

[11] This example is from Schwitzgebel (2011), who uses it in the context of a more general argument to the effect that we're not particularly reliable introspectors, even concerning our present conscious experience.

But my real reason for pessimism about the neo-Cartesian idea that we really *do* learn facts about our own minds with certainty concerns the views in the philosophy of mind that it seems to me to require. In particular, I think it should be most appealing to those who think of mental facts—specifically, facts about consciousness, or seemings, or whatever mental facts get to play the foundational epistemic role of being what we learn with certainty—as both metaphysically and conceptually independent of facts about the rest of the physical world.[12] But for those of us sympathetic to broadly naturalistic conceptions of the mental—such as the interpretivist view sketched in the previous chapter—this response should, I think, seem unattractive. For instance, take the toy Sellarsian (Sellars, 1956) view that for an organism to be in a state of seeming to see blue, it must be in a state that is normally a response to blue things (but which could, e.g., be a response to green things under abnormal lighting). If "I seem to see blue" means "I'm in a state that normally registers the presence of blue things," then you can't be certain that you seem to see blue while at the same time being completely agnostic about the character of the external world; in taking yourself to know with certainty that you seem to see blue, you'd thereby be committed to certainty in the existence of an external world, along with various more specific facts about what it contains (namely, some blue things) and how you relate to it (namely, some of your internal states reliably register the presence of blue things). More generally, if states of consciousness, seeming, and the like, are ultimately functional states of organisms, then it's implausible to claim that we have the kind of skepticism-proof epistemic access to such states necessary for this neo-Cartesian strategy to be plausible. So if you were doubtful that we can know with certainty that the die landed even, you have effectively the same reasons to doubt that we can know with certainty how things seem to us. Essentially, I'm skeptical that we have available a set of concepts or categories for classifying our experiences without making hefty presuppositions about the world beyond our experiences.

As I have no novel arguments to offer concerning these deep issues in the philosophy of mind, I'll turn to a different sort of response to the problem of certainty in Bayesian models of belief. This is the response proposed by Jeffrey (1965). Jeffrey offered a generalization of the Bayesian framework for representing learning. In typical Bayesian models, the input to a learning episode is a proposition (a subset of Ω) that the agent learns with certainty. In Jeffrey's models, the input is not a proposition, but a partition on Ω, along with a set of weights attached to each cell of the partition. The simplest cases

[12] E.g., Chalmers (1996).

are two-cell partitions, though the approach is easily generalized beyond them. Let's return to the case of learning that a die landed even. In an orthodox model, our agent becomes certain that the die landed even, but holds fixed the ratios between the remaining living possibilities. Suppose our Jeffrey agent updates on the partition {even, odd} with the new weights $\left\{\frac{9}{10}, \frac{1}{10}\right\}$. So instead of becoming certain that the die landed even (while holding fixed her opinions *conditional* on the die's landing even), she's becoming confident to degree $\frac{9}{10}$ that the die landed even, while still—as in the orthodox model—holding fixed her opinions conditional on the die landing even (as well as her opinions conditional on the die landing odd). This will give the result that each of {2, 4, 6} will end up with probability $\frac{3}{10}$, and each of {1, 3, 5} will end up with probability $\frac{1}{30}$.

It's not so easy to imagine what sort of learning episode could be fruitfully modeled in the way imagined in the previous paragraph; what sort of experience would make it rational to update one's beliefs by becoming newly confident to degree $\frac{9}{10}$ that a die landed even? The example Jeffrey gave to motivate his approach was of viewing a colored cloth in poor lighting, where based on one's visual experience the cloth might be green, blue, or maybe (though less likely) violet. One can Jeffrey update on the partition {green, blue, violet} with weights reflecting how likely each of these possibilities is in light of one's experience without becoming certain of any of them.

Before saying what I *will* argue for in the remainder of this section, I want to be clear on what I *won't* be defending. I won't be arguing that Jeffrey conditionalization is never or rarely an illuminating framework for modeling learning. I'm happy to concede that it may well be the right tool for many jobs. What I will argue, by contrast, is that modeling learning by Jeffrey conditionalization doesn't really get around the problem of certainty; if you were worried that orthodox Bayesian models of learning were overly idealized because they treat agents as becoming certain of what they learn—when in fact we are not and should not be certain of what we learn—you shouldn't be reassured by the move to Jeffrey-style models. While Jeffrey-style models allow for uncertain learning, they still treat an agent's evidence as—in a very particular sense I'm about to explain—immune to undercutting defeat. And if you were worried about certainty, you should probably be just as worried about immunity to undercutting defeat. So ultimately, the upshot of the discussion will be that the distinction between strict and Jeffrey-style models of learning is less philosophically significant than one might have initially thought. In particular, neither framework should give comfort to the ambitious modeler who yearns for a single model that would capture how a rational agent would update her beliefs over time in response to arbitrary new information.

The limitations of Jeffrey conditionalization I'll discuss are not novel—they're adapted from Christensen (1992) and Weisberg (2009, 2015). But both of those writers treat these limitations as deep problems for the Jeffrey conditionalization framework. By contrast, the lesson I'll draw is that modelers using the Bayesian framework to represent learning—whether traditional or Jeffrey-style—should be modest in their aspirations.

Before discussing immunity to undercutting defeat, I want to contrast it with the more familiar idea of indefeasibility.[13] What is indefeasibility, and how does it relate to certainty? Certainty concerns an agent's beliefs at a time—if an agent is certain that P, then she is maximally confident that P, and her practical decisions are insensitive to costs of actions that would only accrue if P were false. Indefeasibility concerns an agent's dispositions to change her attitudes—or not—in response to new evidence. To a first approximation, we might say that an agent indefeasibly holds an attitude if they not only currently hold it, but they're also committed to continuing to hold it no matter what future evidence they get.[14]

Consider an orthodox (not Jeffrey-style) Bayesian model with discrete time where an agent's beliefs at t_{n+1} are obtained by conditionalizing her beliefs from t_n on whatever new evidence she learns at t_{n+1}. In such models, if an agent learns some evidence E at any time, she remains certain of E for all future times. So in such models, certainty and indefeasibility are closely linked; once you've learned something, it remains indefeasibly certain going forward. But in other models—even other orthodox models—beliefs held with certainty need not be indefeasible. For instance, in models where an agent's beliefs at a time t_n are obtained by starting with a constant "ur-prior" and conditionalizing that ur-prior on the agent's total evidence at t_n (rather than conditionalizing it on the incremental evidence gained from t_{n-1} to t_n), then there's no barrier to some proposition being entailed by the agent's evidence—and thus certain—at t_n, but not entailed by their different body of evidence, and thus uncertain, at t_{n+1}.[15] In effect, such models allow for forgetting.[16]

And if orthodox models can allow for our certainty in evidence propositions to be defeasible, it's pretty obvious that Jeffrey-style models can also allow

[13] This is somewhat stipulative. Some people use "indefeasibility" to mean what I'm calling "immunity to undercutting defeat." I don't care too much about the labels, so long as we recognize the distinction under some name or another.

[14] So as I'm using the terms, it's not only beliefs held with certainty that can be indefeasible. If one thought, for instance, that there was no possible evidence for or against various claims—that the universe was created by a loving God, or by a clever simulator—one might hold indefeasibly *agnostic* attitudes towards those claims. Nevertheless, it's typical to focus on the case of indefeasible certainties.

[15] See Meacham (2016) for a recent discussion of ur-prior conditionalization.

[16] See also Williamson (2000, ch. 10) who offers such a model, and emphasizes its ability to accommodate forgetting.

for our high but uncertain confidence in some cell of an evidence partition to be defeasible too. If I Jeffrey update on the partition {E, ~E} with weights {.99, .01}, nothing stops me from Jeffrey updating on that same partition at the next moment with the flipped weights {.01, .99}.

So neither traditional nor Jeffrey-style models seem forced to regard our attitudes towards what we learn as indefeasible. Nevertheless, I claim that Jeffrey conditionalization *does* face a prima facie problem with being forced to regard our attitudes towards what we learn as immune to undercutting defeat. So what is immunity to undercutting defeat, and how does it differ from indefeasibility?

To explain immunity to undercutting defeat, we have to start with undercutting defeat, and to explain undercutting defeat, we should start with defeat more generally.[17] While different formal frameworks will spell it out differently, the basic idea is that defeating evidence counts against a hypothesis; defeating evidence is disconfirming evidence.

With this generic notion of defeat in hand, we can now distinguish between "rebutting" and "undercutting" defeat. I'll start with an intuitive sketch of the distinction, then offer some examples, and then explain how the distinction can be captured in either strict or Jeffrey-style models of learning, ultimately building up to the idea that *neither* framework can capture the idea of an agent's evidence itself being vulnerable to undercutting defeat.

The basic idea is that rebutting defeat directly counts against a hypothesis, while undercutting defeat merely weakens some support the hypothesis already had. So if I look at one clock that says it's 4:00, and then I look at a second that says it's 5:00, my viewing of the second clock has provided rebutting defeat for the claim that it's 4:00. By contrast, if instead of looking at a second clock, someone I completely trust tells me that the first clock is stopped at 4:00, I've gotten undercutting defeat for the claim that it's 4:00. Kotzen (2019) provides a thorough account of how the distinction can be modeled in the Bayesian (strict conditionalization) framework. As a first pass, R provides rebutting defeat for H when $P(H|R) < P(H)$. By contrast, U undercuts E's support for H when the following three conditions hold:

1. $P(H|E) > P(H)$.
2. $P(H|U \& E) = P(H)$.
3. $P(H|U) = P(H)$.

[17] In using "defeat" as an umbrella term, and then distinguishing "undercutting" from "rebutting" defeat, I'm adopting the terminology of Pollock (1987). Elsewhere in the literature sometimes "undermining" is used instead of "undercutting" and/or "opposing" is used instead of "rebutting."

The first condition captures the idea that E supports H. The second captures the idea that U defeats this support—once U is there to undercut that support, H's probability reverts to its prior. The third condition captures the idea that U *doesn't* directly count against H. In the taxonomy of Kotzen (2019) these three conditions describe a kind of paradigmatic case of total undercutting defeat, but we can have partial and/or mixed cases as well, such as when the probability of H given E and U doesn't make it all the way back down to the prior probability of H, but is still lower than the probability of H given E alone.

Let's return to the example of the clock, letting H = "it's 4:00," E = "clock says it's 4:00," and U = "The clock is stuck at 4:00." Here, it's plausible that all three conditions are met, at least assuming that I start off rationally uncertain of the time, and disposed to trust clocks (albeit not completely). The probability that it's 4:00, given that the clock said so, is greater than the prior probability that it's 4:00. But the probability that it's 4:00, given that a clock that is stuck at "4:00" displays "4:00," is identical to the prior probability that it's 4:00. Still, the fact that the clock is stuck at 4:00 doesn't count *against* the claim that it's 4:00.

Can we represent undercutting defeat in the Jeffrey framework, where the inputs to learning episodes aren't propositions but input partitions along with associated (non-extreme) weights? We can. While I'll provide a simple worked example in the next paragraph, intuitively, not much about the previous example would've changed if instead of strictly conditionalizing on the claim that the clock read 4:00, and then on the claim that it was stopped, we instead performed Jeffrey updates on two-cell partitions and became highly confident but not certain first that the clock read 4:00, and then that it was stopped.

Let's say that a case of undercutting defeat in the Jeffrey framework is one where one Jeffrey update J_2 undercuts the support J_1 provides for H just in case:

1. H's probability after J_1 is higher than its prior.
2. H's probability after J_1 followed by J_2 is back to its prior (or at least closer to its prior than it was after J_1).
3. H's probability after only J_2 is identical to its prior.

To see that such a case is possible, let's return to an example involving a die. Let H = the die lands even, let J_1 be an update on the two-cell partition {the die lands > 3, the die lands ≤ 3} with weights {0.9, 0.1}, and let J_2 be an update on the two-cell partition {the die lands between 2 and 5, the die lands either 1 or 6} again with weights {0.9, 0.1}. The prior probability that the die lands even is $\frac{1}{2}$. When we undergo J_1 and become highly confident that the die landed greater than 3, we think it's probably even (since we're now confident that the die landed 4, 5, or 6, and two out of those three possibilities are ones where it landed even.) But when we become highly confident that it landed on neither 1 nor 6, then we're back to almost where we started concerning whether the die landed even; becoming confident that it didn't land on 1 or 6 undercuts the support that we had for thinking it was even. But if we'd become confident that it landed between 2 and 5 without first becoming confident it landed greater than 3, then our confidence that it landed even would've remained unchanged, at $\frac{1}{2}$; becoming confident that the die didn't land 1 or 6 doesn't *rebut* the hypothesis that the die landed even.[18]

The (putative) difficulty for Bayesian accounts of undercutting defeat is the following: just as one can undermine support for a hypothesis by attacking the link between one's evidence and the hypothesis, one can also undermine support for a hypothesis by attacking the evidence itself. So not all undercutting defeat can be understood as involving the accumulation of evidence, where the final, total body of evidence fails to support some hypothesis that was supported by the initial, smaller body of evidence. Arguments of this general form have been given by David Christensen (1992), Jonathan Weisberg (2009), and James Pryor (2013).

To see how this worry gets off the ground, let's use a new example. Suppose you read in a newspaper that the majority of Salemites intend to vote "Yes" on proposition 17.[19] As a result, you come to believe that proposition 17 will pass. Later, however, you read that the polls on which the first article was based come from a source with a flawed methodology—their results are not likely to be representative of what the people of Salem, on the whole, intend. Intuitively, this is a case of undercutting defeat. But we need to take care about how we represent it if we're going to capture this thought. If we represent the initial example as one in which you strictly conditionalize on the evidence that the majority of Salemites intend to vote "Yes" on proposition 17, then we don't get the desired result. Let:

[18] The calculation for the probability that the die is even after J_1 is $\frac{9}{10} \cdot \frac{2}{3} + \frac{1}{10} \cdot \frac{1}{3} = \frac{19}{30}$. But after J_1 *and* J_2, it's not all the way back down to $\frac{1}{2}$, but it's close, with the calculation being $\frac{9}{10} \cdot \frac{1}{2} + \frac{1}{10} \cdot \frac{9}{10} = \frac{27}{50}$. The reason why we don't get all the way back down to $\frac{1}{2}$ after J_2 is that even though most of the probability mass is concentrated on 4 and 5, 6 is still more likely than 1 (as well as more likely than 2 and 3).

[19] I used this example in my (2017).

Intention = The majority of Salemites intend to vote "Yes" on proposition 17.
Pass = Proposition 17 will pass.
FlawedSource = A second story reported that the polls relied on by the first story come from a source with a flawed methodology.

What we'd *like* to say, if we're going to represent the case as an instance of undercutting defeat in the framework of strict conditionalization, would be that the following three conditions hold:

1. P(*Pass* | *Intention*) > P(*Pass*).
2. P(*Pass* | *Intention* & *FlawedSource*) = P(*Pass*).
3. P(*Pass* | *FlawedSource*) = P(*Pass*).

But the second condition *does not* hold. If the majority of Salemites intend to vote "Yes," then regardless of how the polls were conducted, the proposition will probably pass. That is, P(*Pass* | *Intention* & *FlawedSource*) > P(*Pass*).

The basic problem is that if we treat the evidence provided by the first story as *Intention*, then that evidence is not defeated by the story about the flawed polling methodology. Rather, if we are to accommodate the possibility of undercutting defeat via that story, it looks like we need to treat the evidence provided by the first story as something like *Polls*:

Polls = A poll reported that the majority of Salemites intend to vote "Yes."

Only if something like *Polls* was the evidence that the first story provided can we explain why the story about the flawed methodology defeats your support for thinking that the proposition will pass.

Crucially, the difference between strict and Jeffrey conditionalization isn't playing any important role here. Even if we represent the case as one with a Jeffrey update on the partition {*Intention*, ~*Intention*}, with most of the weight on the first cell, we still can't capture the idea that performing a second Jeffrey update on the partition {*FlawedSource*, ~*FlawedSource*} undercuts the support the first update provided to *Pass*. In order to allow that *FlawedSource* undercuts the support that *Intention* provides for *Pass*, *FlawedSource*, and *Pass* must start out probabilistically independent of one another, but must become probabilistically dependent (in particular, they must become negatively relevant to one another) after Jeffrey conditionalizing on the partition [*Intention*, ~*Intention*]. But Jeffrey conditionalization cannot induce this sort of probabilistic dependence. This is a consequence of the fact that Jeffrey conditionalizing leaves probabilities *conditional* on each cell of its update partition untouched. More intuitively, conditionalizing on *Intention*—whether strictly, or Jeffrey-style—does not give us a way to represent that the support for *Intention* depends on beliefs about the polls, and so does not give us a way to represent that the support for intention can be defeated by information about polls.[20]

So far this might not seem like a problem at all. What's wrong with treating the evidence provided by the first story as *Polls*? Or alternatively, with representing it as a Jeffrey update on {*Polls*, ~*Polls*}? The danger is that just as *Intention* can be undermined, so too can *Polls*. For instance, suppose the second story is just a correction—it reports that the first story misreported the poll results. If this undercutting defeater is to be accommodated, the evidence will need to concern, not what the polls said, but what the newspaper said the polls said. More generally, the problem is that it's very hard to find a good candidate for the sort of proposition whose probability should be *directly* affected—whether by going all the way to certainty, or just by going higher—by a learning experience, no matter what the agent's background beliefs, what defeat possibilities are salient, or anything else like that.[21]

If I were more optimistic about the neo-Cartesian projects mentioned earlier, I'd probably say that the buck stops at *experience*.[22] A rational agent can be modeled as conditionalizing—maybe strictly, maybe Jeffrey-style—on claims about her experiences. And such claims will be immune to undercutting defeat. So once we represent the evidence as something like: "I'm having an experience as of the newspaper reporting certain poll results," no further retreat will be necessary. But as already discussed, I'm skeptical that a suitably naturalistic conception of experience—or consciousness, or seemings—can be made to bear this much philosophical weight. I have a much better grip on how, in particular cases, it seems rational for agents to change their beliefs in worldly propositions than I do on the ineffable experience propositions the neo-Cartesian ends up positing as the putatively immediate drivers of such changes.[23]

[20] See Weisberg (2009) for a fuller explanation.

[21] See Christensen (1992), who puts the point in terms of a conflict between confirmation holism and Bayesian epistemology.

[22] Or maybe consciousness, or seemings, if any of these are different.

[23] Schwarz (2018) provides an account of rational updating meant to avoid these worries, in which the propositions directly learned in experience are "virtual," non-worldly propositions, which can be understood only in terms of their probabilistic relations to more familiar worldly propositions. His account is ingenious, but unusual, and fully engaging with it here would take us far afield. In brief, I think the model he proposes is a fruitful one, but I suspect that if interpreted with the ambition that I think Schwarz intends, it will still run into the familiar "problem of new theories," introduced

To sum up, the challenge to Bayesian accounts of undercutting defeat is as follows. If the Bayesian is too generous about what our evidence is—e.g., if she thinks that our evidence in the cases I have been discussing is *Intention*, or if she represents our evidence as a Jeffrey update on {*Intention*, ~*Intention*}—then she will not be able to account for certain cases of undercutting defeat. But there's no way to avoid being too generous; however the Bayseian characterizes our evidence, it will be immune to certain intuitive instances of undercutting defeat. But just as nothing is certain, so also nothing is immune to undercutting defeat.

A modest modeler might think these difficulties call for supplementing the Bayesian framework with an alternative one to accommodate undercutting defeat; she'll have no principled objections to the idea that not all epistemological phenomena can be captured in a single modeling framework. However, there are reasons to be unhappy with the idea that an alternative framework, or even an extension of the Bayesian approaches we've been discussing so far, is the right tool for this job. First, alternative frameworks available so far run into similar troubles in accounting for undercutting defeat.[24] Second, the Bayesian account of undercutting handles some cases—e.g., the first version of the newspaper case—extremely nicely, and these cases do not seem all that different from the cases where the Bayesian account seems to falter. If we rest content with the Bayesian treatment of the original case involving a story about polls and a story about flawed polling methodology, but look for some

by Glymour (1980). Schwarz's account requires that there be a single space Ω such that all of what an agent learns can be modeled as shifting her probability distribution over the subsets of Ω. This view has a hard time accommodating learning episodes in which what one learns is that there are possibilities of which one was previously unaware. A modest modeler can wait to hear which such possibilities are being discussed, and then include them in her model from the start. An ambitious modeler needs to think that all such possibilities could be included from the outset. But it's far from clear to me that this could be done, even in principle. The notion of a possibility might be indefinitely extensible, in a sense analogous to what some philosophers of mathematics have claimed on behalf of the notion of number, or set. Maybe, for any set of possibilities, there is a larger one. Or for any set of distinctions among ways the world might be, there is a more fine-grained set of distinctions. There has been a great deal of debate in the philosophy of logic and mathematics over whether the notion of absolutely unrestricted quantification—quantification over absolutely everything—makes sense. (See Uzquiano and Rayo (2006) for a volume on the topic.) And at least some of the reasons for thinking it doesn't seem like they would carry over to provide reasons for doubt about the existence of a set not of all things, but of all possibilities in which a rational agent might invest credence. E.g., Agustín Rayo writes that absolutely unrestricted quantification "would require a final answer to what counts as a possible system of compositional representation. And I see no *prima facie* reason to think that our notion of representation (and our notion of linguistic representation, in particular) are constrained enough for this question to have a definite answer" (2013a, p. 29). If he's right, then it's similarly plausible that there's no definite answer to what counts as a possibility—for any system of representation that allows one to generate a space of possibilities Ω over which credences might be defined, there would be a richer system of representation that would allow for distinctions and possibilities not recognized by the previous one.

[24] See Weisberg (2015).

alternative framework to handle the variations on the original case, we risk ending up with an account of defeat that looks oddly disjunctive. At least to me, it seems that the sort of defeat provided by the story about the flawed methodology is not all that different from the sort of defeat provided by the story about the error in reporting the poll results. It would be nice to be able to treat both of these cases as having a similar formal structure, so if we're going to accept a Bayesian model of the first case, (which is hard to resist) it would be nice to have a Bayesian model of the second as well. Ultimately, I will argue that such an account can easily be had, so long as we're modest about model selection.

Let's take stock. I started the chapter by noting a tension between the intuitive idea that certainty is unattainable, and the fact that formal models of belief seem to represent agents as certain of a great deal. I suggested the typical view about this tension is that the certainties posited by formal models of belief are idealizations. This suggested a project of trying to say something about the non-idealized truth that these idealized models are approximating. I've been reviewing some difficulties for that project; it's very hard to say what the strict, non-idealized truth is about what sort of information we learn from experience. In the next section, I'll try to show how we can decline to take up that project, resting content with the idea that the best we can do in describing rational belief and how it changes over time is to use a variety of idealized models, none of which will be universal in scope.

4.3 Certainty and Modesty

My aim in this section is to describe how a Bayesian who is modest about model selection can think about the difficulties associated with certainty and undercutting defeat described in the previous section.

Consider the game of rock paper scissors with one slight modification. Instead of players making their selections simultaneously, one player goes first and the other player goes second, with the knowledge of the first player's choice. In this variant, the second player can obviously guarantee a win. I claim that if our Bayesian is modest about model selection, her position is analogous to that of the second player in this game; given a decision or learning problem described in natural language, she can tailor her formal model of the problem to make sure that whatever must be uncertain, is uncertain, and whatever must be vulnerable to undercutting defeat, is vulnerable to undercutting defeat. She only gets into trouble if she takes on further, more ambitious commitments, to the effect that her model can adequately capture not only the problem as it was initially posed, but also arbitrary elaborations of that problem. This would be analogous to a game of rock paper scissors in which one player makes her choice, the second makes hers with knowledge of the first, but then the first is allowed to revise her choice in light of the second's.

Let's unpack the analogy. Consider a two-player game in which the first player tells a story in natural language about some agent(s) facing some practical or theoretical problem in which uncertainty and/or undercutting defeat intuitively play an important role. The second player, the Bayesian modeler, responds by using her framework to build a formal model meant to generate predictions or recommendations (depending on whether she's doing descriptive or normative modeling) concerning how the agent(s) will act and/or update their beliefs.[25] You're the referee. You'll judge that the second player wins if the predictions/recommendations strike you as sensible ones.[26] If the predictions/recommendations strike you as bad ones, you'll award the victory to the first player.

Nothing in the previous section, I suggest, should lead us to worry that the second player will have a hard time winning this game. If the first player's story is one where it's important that some fact is uncertain, then the second player can make sure to build a model in which that fact is uncertain. For instance, if the first player tells a story where an agent is faced with the question of whether to buy flood insurance, then the second player can model the information of the agent in the story with a set of possibilities that includes ones where the agent's house is flooded. And if the story is one in which undercutting defeat plays an important role—the story involves two learning episodes in which, intuitively, the first supports some hypothesis, and then the second undercuts that support—then the second player can accommodate this too. For instance, suppose the story told by the first player is one of the ones from the previous section, with a character who first reads an article about Salemites intending to vote for proposition 17, and then reads a second article alleging that the polls relied on by the first article were flawed. In that case, our second player can make sure to model the effect on the agent of reading the first article as an update on something like *Polls*, rather than something like *Intention*.

So why, again, were we supposed to worry that the Bayesian has a problem with certainty and/or undercutting defeat—what's the game that the Bayesian

[25] Here I'm inspired by Titelbaum (2012), who thinks of the objects modeled by the Bayesian framework he offers as "stories."

[26] Alternatively the story is about a real life, concrete situation, and the task is a descriptive one, then we can say the second player wins if her predictions are close enough to what actually happens.

modeler *can't* win? Here's how it's played. The first player tells the beginnings of a story. The second player provides a formal model of the story thus far. The first player then continues the story. The second player must then develop her model to represent the continuation of the story, but without changing any aspects of her model of the first part of the story. So, if the model provided by the second player involved a possibility space Ω, and an agent learning some subset E of Ω, then the model of the full story must still involve that same space Ω, and that same update on E—it can just have more that happens afterwards.

Why is this game so hard? At least if the second player builds the kind of humanly tractable, small-world models familiar from economics and decision theory, then the first player has a winning strategy. She can look for some possibilities not included in the original model, and then develop her story in such a way that this omission is a fatal flaw. For instance, imagine the game goes as follows:

Player 1: An agent is making some bets on the outcome of rolls of a die. She learns that the die landed even. Will/should she accept a bet that costs $0.2, and plays $1 if it landed on 6?

Player 2: I'll model her information with the possibilities {1, 2, 3, 4, 5, 6}, and equal probability assigned to each, and then her new evidence as an update on {2, 4, 6}, with the result that 6 ends up with probability $\frac{1}{3}$. Given natural assumptions about her utility function, accepting the bet has greater expected utility than rejecting it, so I predict/recommend she'll accept it.

Player 1: Wait, before she has a chance to accept the bet, she sees another die on a table nearby—not the die whose outcome she's betting on—and on that die all three of the showing faces are 4. How, if at all, should this affect her willingness to accept the bet?

It's pretty clear that Player 2 can't give an adequate answer to this question if she has to stick with the model she offered in response to the first part of Player 1's story. That model treated it as certain that the agent was betting on the toss of a normal, 6-sided die. But if we're going to model the epistemic significance of seeing a trick die, we need a more complex model with a larger event space.

I expect you'll agree both that the first game seems feasible enough for Player 2 to do well at, and that no human could do well in the second game as Player 2. Someone who is modest about model selection is happy enough to be able to consistently win the first game in the position of Player 2; to the extent that there are stories she can't handle, she'll try to expand her toolkit of models so as to be able to better understand a wider variety of scenarios. By contrast, while someone who's ambitious about model selection doesn't aspire to be able to win the second game in the position of Player 2 *herself*—even if she's ambitious, she'll know her limits—she *does* want to be able to say something about the modeling framework a less limited creature could use to win the second game. And this is where worries about certainty and undercutting defeat come in. Even someone who could intellectually grasp or somehow write down a truly vast state space Ω—e.g., somebody who'd already included possibilities where the die is unfair in response to the first part of Player 1's story—will *still* lose to a clever opponent, if she's building the sorts of Bayesian models we've been discussing. That is, Player 1 can tell a story involving a learning episode, and then however Player 2 models that learning episode, Player 1 can develop the story in such a way that Player 2's model will be unable to capture what intuitively looks like a case of undercutting defeat.

This, at any rate, is how I interpret authors who treat the considerations discussed in the previous section as a challenge to Bayesian frameworks for modeling rational belief.[27] They take it that epistemologists should be trying to say something about what a winning strategy for Player 2 in the second game would look like, even if they know *we'll* never be able to execute that strategy. And so if it looks like Bayesian frameworks for modeling rational belief and action even in principle can't provide someone with a winning strategy for playing the second game, we should be on the lookout for alternatives that can.

By contrast, someone who is modest about model selection can be agnostic about whether there *is* a winning strategy for Player 2 in the second game, at least if Player 2 has to stick to roughly the level of description at which epistemologists, economists, and decision theorists tend to work. It's easier to imagine a winning strategy if we can describe a rational agent at the cell-by-cell (or circuit-by-circuit) level, from which various claims about how she'd respond to novel stimuli would follow. But if we're building models of a rational agent using folk psychological categories or their mathematically regimented cousins, perhaps winning the first game is the best that can be done. Of course, if some alternative comes along that looks like it *could* provide a computationally unlimited agent with the means to win even the second game, sticking to models using categories at roughly the same level of description as folk psychology and/or decision theory, that would be exciting and worth exploring. But in the meantime, the modest modeler is happy to

[27] Such as Christensen (1992), Weisberg (2009, 2015), Pryor (2013), and Schwarz (2018).

focus on improving her play at the first game, and so needn't see any pressing reason—at least any pressing reason coming from problems with certainty and undercutting defeat—to look for alternatives to the Bayesian frameworks that we've encountered thus far.

4.4 Whither Normativity?

I hope I've made it clear why someone who's committed to using the Bayesian framework for representing rational belief, but is modest about model selection within that framework, needn't be particularly concerned about the way in which Bayesian models inevitably treat some claims as certain and/or immune to undercutting defeat. Now I'll address a worry to the effect that modesty in model selection sits uneasily with the first-personal practice of using normative models to decide what to do or think.

Imagine you're faced with some problem where constructing a decision theoretic model seems like a fruitful way of trying to decide what to do or think. This book is being written in the midst of a global pandemic that provides plenty of opportunities for applied decision theory. If I've been trying hard to avoid becoming infected, how—if at all—should my behavior change once I've been vaccinated? It's natural to think about this sort of question by trying to come up with estimates for my chance of being infected—and then conditional on that, my chance of suffering further bad outcomes—if I engage in various behaviors (domestic travel? international travel? indoor dining?), and to then ask how those probabilities change conditional on my being vaccinated, and then on top of all that trying to come up with a utility function that will represent how strongly I care about avoiding various bad outcomes and achieving various good ones, and then picking the expected utility maximizing behaviors. OK maybe that's not *really* all that natural. But even if few people actually do all that—I didn't—I claim it *is* natural to approach these questions by thinking about what the result would likely be if one *did* do all that. So let's imagine the situation of someone who has constructed a Bayesian decision theoretic model of her practical situation, with the aim of deciding what to do in light of her values and evidence. Why might she be uneasy about acting on the basis of her model's recommendations?

One natural thought is that if you know there are a variety of ways your practical situation could be modeled—each of which strikes you as prima facie reasonable—*and* that they give different recommendations concerning whether some course of action is utility maximizing, or whether some learning episode provides evidence for or against some hypothesis, then you'll be uncomfortable relying on any of those models in deciding what to do or think.[28] True. But this strikes me as an unavoidable problem, and one that arises for the ambitious modeler in essentially the same way. Whether or not there's in *principle* a complete, non-idealized, grand-world decision theoretic model of your situation somewhere in Plato's heaven, in practice we must rely on incomplete, idealized, small-world models in deciding what to do. And when the small-world models that we come up with differ on whether, given our current evidence and values, it's a good idea to eat at an indoor restaurant (for example), there's not much we can do to satisfyingly resolve that question.[29] There exist hard cases, where normative modeling doesn't generate determinate recommendations. But this strikes me as an unavoidable feature of normative theorizing, and one that pops up in a variety of guises even among those who don't approach these questions through the lens of modeling.[30]

Perhaps the issue is that the ambitious modeler can at least have a clear idea of *what it is* we're uncertain about in these hard cases: we're uncertain as to just which course of action is represented as most choiceworthy in the grand, not-humanly-graspable model up in Plato's heaven. But the modest modeler, who's agnostic or skeptical about the existence of such a model, can't understand her uncertainty in these terms. So how *can* the modest modeler think about just what it is she's uncertain about, when different models give different recommendations concerning what she should do or think?

While this might seem like a distinctively hard problem for the normative modeler, I think we can ultimately understand such uncertainty along very similar lines to how we can understand uncertainty about matters of descriptive fact. Consider an economist who has constructed a variety of models of some market, each of which strikes her as prima facie sensible, and which make very different predictions for how the market would respond to some intervention. Perhaps she's modeling a labor market, and trying to predict how it would respond to an increase in the minimum wage.[31] If she's a modest

[28] I skip straight to the case where the available models disagree because it's harder than the case where the recommendations your normative models generate are highly robust to different prima facie reasonable modeling choices.

[29] Unless, perhaps, there's some prospect of their agreeing once we gather more evidence, and it's feasible for us to gather that evidence.

[30] See, e.g., Smith (1988) and Srinivasan (2015), both of whom argue that there are principled limits to the degree to which ethics can be operationalized. Or see the vast literature on moral dilemmas growing out of Marcus (1980).

[31] The variety of models of labor markets is an example Rodrik (2015) frequently returns to. As he emphasizes, while there are familiar models which predict minimum wages should tend to increase unemployment, there are also monopsony models where minimum wages have the opposite effect. It's an empirical question, he argues, which models best fit which particular concrete situations.

modeler, she won't think of her uncertainty as wondering which of these models has most in common with the *true* economic model of the situation; she doesn't think there is a true economic model of the situation. Rather, her situation is more like that of someone who thinks the fundamental truth is some version of quantum mechanics, and is wondering which of some family of classical models provides the best approximation of certain aspects of some fundamentally quantum mechanical system.[32]

So what would a normative analog of this stance look like? You'd have to think that while the fundamental, non-idealized normative truths can't be represented in any of *our* modeling frameworks, they can be represented in some other one—something like a normative analog of fundamental physics. Maybe it would be a framework for describing which arrangements of fundamentalia instantiate which species of value to which degrees, or something along those lines—I don't want to make any strong assumptions about what sorts of categories it would use.[33] Uncertainty about which of various tractable, small-world normative models of my practical situation is best could amount to uncertainty about which of those models best approximates the fundamental normative truth about the situation, which would be statable only in some vocabulary to which we sadly lack access.

This normative analog of fundamental physics that I'm imagining is a bit like the category of the "protophenomenal," proposed by Chalmers (2013). His idea is that while we have good reason to think phenomenal facts—facts about consciousness—can't be grounded in the facts of fundamental physics, the ontology of the phenomenal is too varied and diverse to plausibly be fundamental. Rather, it's more attractive to think there is some category of "protophenomenal" facts, which stand to the phenomenal facts we know and love—facts about what it's like to see red, or to hear a trumpet—as facts about fundamental physics stand to facts about macroscopic physical objects.[34] I'm suggesting a similar analogy, where there are fundamental protonormative facts, and we can think of our humanly tractable normative modeling frameworks as letting us build fruitful models of the more fundamental protonormative facts.

[32] As in David Wallace's example of the solar system, mentioned in Chapter 2.

[33] Just as it would be unwise for people at our stage of intellectual development to make strong assumptions about which categories truly fundamental *physics* will end up using. Particles? Fields? Strings? Something else? I don't know. The view I'm sketching suggests we should be similarly agnostic about which categories would be used in stating the fundamental truths which our ordinary talk of "reasonable," "ought," "good," and the like provides serviceable models of.

[34] To be clear, I'm not endorsing Chalmers' views about this matter; I'm more optimistic than he is about the prospects for physicalism in the philosophy of mind.

While this view may sound odd, I think it actually has a good deal of independent appeal. The idea that even the most ambitious theories in normative ethics—at least the ones we have so far—are at best in the business of providing serviceable models of more fundamental proto-normative truths becomes more plausible once we think about cases of normative vagueness.[35] Imagine a utilitarian who thinks the fundamental normative truth is simple—more pleasure is good, more pain is bad, the rest is just details. While perhaps an implausible view, it at least seems to have the virtues of ontological parsimony, generality, and precision. But not so fast. Consider the question about whether it's bad that some worm be squished.[36] If worms feel pain then it would be bad that the worm be squished; being squished hurts. But whatever state worms are in when they're getting squished is, plausibly, a borderline case of pain. If you disagree or are skeptical, I expect you'll agree that there's *some* location on the tree of life where the claim is plausible. Suppose you also think that fundamental normative truths can't be vague.[37] It's natural at this point to think that to *really* capture the normative truth about the situation, we need to start using categories other than pain. We need to talk about whatever more fundamental facts underlie facts about pain, and what *their* normative significance is. E.g., to what degree is it bad that a creature be in a state with such-and-such functional features? Or something along those lines. The point is that if you don't want the fundamental normative truth to be vague, then you had better think the normative ethical theories we have on the table—theories that talk about things like pain and pleasure, or intentional actions, or consent, or virtue—are not candidates for statements of the fundamental normative truth, since the prospects for eliminating vagueness from such theories are very poor.

[35] The argument I'm about to offer is inspired by Schoenfield (2016), though I end up in a very different place. In effect, she argues that (1) facts about which actions are permissible are vague, so (2) if the fundamental moral truth includes facts about which actions are permissible, then the fundamental moral truth is vague. And moreover, (3) vagueness in fundamental moral truth must be "ontic" vagueness rather than epistemic or representational vagueness. I agree with all three of these claims, but am suggesting that rather than accepting that the fundamental moral truth includes facts about permissibility and performing the modus ponens to conclude that the fundamental moral truth is vague, it's more attractive to perform modus tollens, and to conclude that whatever the fundamental moral truths are, they must be some non-vague truths of which our talk of permissibility provides serviceable models.

[36] I pick a worm because they strike me as a plausible case of an animal where it's vague whether or not they can feel pain. While I grant there are views in the philosophy of mind on which this is *not* plausible—views where facts about pain are themselves fundamental—if you think pain is a kind of functional state of an organism, then presumably you should also think there are cases of organisms in states that play some but not all of the characteristic functional roles of pain.

[37] E.g., because you think all vagueness is in our concepts, rather than in the world, as argued by Sainsbury (1995).

That was a roundabout way of explaining how a modest modeler might go ahead using normative models in a first-person, deliberation-guiding way, all the while knowing that the models she's using are idealized, and that other models might well disagree. She can think of the models she uses to generate recommendations concerning what she ought to think or do as her best attempt to approximate some more fundamental normative (or proto-normative) truths, statable only using some set of categories that she can't herself imagine. While this might sound like a strange view, I've tried to show how it can be motivated by thinking about vagueness, via a route that doesn't appeal to the particular perspective on idealization and modeling that I've been urging in this book.

This might seem like an approach only a robust realist in metaethics could take, and I admit that, prima facie, the idea of proto-normative truths statable only in some vocabulary that we haven't yet conceived has a realist flavor; if normative facts are in any sense something we construct or project onto the world, rather than something "already" out there, shouldn't they be more epistemically accessible than that?[38] While I'm sure there are popular views in metaethics that are incompatible with the picture I've just sketched, I don't think it's true that only a robust realist could find it attractive. Quasi-realists are experienced at taking distinctively realist-sounding claims and interpreting them as expressions of first-order normative commitments or practices, and I don't see why someone who's optimistic about other applications of that strategy should be especially skeptical about its use here.[39]

[38] The idea that normative facts are more epistemically accessible if some variety of antirealism is true is a familiar, albeit contested one. See Street (2006).

[39] E.g., Simon Blackburn (Blackburn, 1993, ch. 1, esp. p. 25) takes the example of a claim to the effect that bivalence holds in some domain and suggests that a quasi-realist like him can interpret such a claim as the expression of a regulative ideal, roughly to the effect that in hard cases we should keep looking for reasons rather than accepting any cases as ones where there's no right answer. I suspect he could say the same thing here; rather than treating the question of whether its bad that fish be squished as irresolvably vague, he could endorse a regulative ideal of looking for further distinctions or categories that would resolve the question, and could treat the claim that there *are* such further distinctions or categories as an expression of that ideal.

5
Belief and Credence

Recent years have seen an explosion of interest in questions about the relationship between qualitative epistemological concepts like belief and knowledge, and quantitative ones like credence or probability. Much of the literature has been framed as addressing questions about the relative ontological priority of two classes of mental state; is belief reducible to credence, or is credence reducible to belief, or is neither attitude reducible to the other?[1] In this chapter I'll reformulate such questions in the language of model-building. While I won't exactly defend a position on the reoriented debate, I will argue that much of the literature goes wrong in presupposing what we rejected in the previous chapter; that the Bayesian modeler must be ambitious, and attribute a single, domain-general set of credences to any given agent. Once we allow that someone using broadly Bayesian models of thought and choice will rightly use different models to capture different aspects of an agent's cognitive life, and that each of these models will treat a great deal as certain, many of the putative reasons to posit belief as a fundamentally different kind of attitude from credence look less persuasive.

In particular, I'll take aim at two main targets. In §5.2, I'll criticize the idea that we need to appeal to a folk psychological, non-decision-theoretic notion of belief to explain how computationally limited creatures like us manage complexity. And in §5.3, I'll criticize the idea that we need to appeal to a folk psychological, non-decision theoretic conception of belief to explain certain normative phenomena, such as when we are justified in punishing and blaming. But first, in §5.1, some stage setting.

5.1 Views about Belief and Credence as Views about What Modeling Frameworks Can Do

Back in Chapter 3 we encountered the idea that the interpretivist tradition in the philosophy of mind could be fruitfully read through the lens of modeling.

[1] See Jackson (2020) for a helpful survey.

Idealization in Epistemology: A Modest Modeling Approach. Daniel Greco, Oxford University Press. © Daniel Greco 2023.
DOI: 10.1093/oso/9780198860556.003.0006

Rather than talking about how an ideal interpreter of a certain sort would interpret people, we can instead talk about what sorts of modeling frameworks could be used to construct good models of people. Rather than saying: "an ideal interpreter would interpret Karl as having beliefs *B* and desires *D*," we can instead say: "the best folk-psychological model of Karl is one in which he has beliefs *B* and desires *D*." Likewise, of course, with credences. The interpretivist who's taken the modeling turn can say that for Karl to have credence *n* that P is for the best Bayesian model of Karl to be one in which he's represented as having credence *n* that P.

Assuming we find this lens a fruitful one, how should we interpret questions about the relative ontological priority of belief and credence? Jackson (2020) provides a flow-chart taxonomy with five main positions: two versions of eliminativism (one which says beliefs exist and credences don't, and the other which says the reverse), and three views on which both credences and beliefs exist, but which differ on which is more fundamental ("credences," "beliefs," and "neither" being the three options). How might we state analogous positions using the ideology of modeling?

Eliminativist positions are perhaps the easiest. A belief-eliminativist, such as Jeffrey (1970), thinks that any phenomenon you might be tempted to explain using folk psychological models, you can explain better using decision theoretic ones. Just as Copernican models of the solar system rendered Ptolemaic models obsolete, so has decision theory, with its credences and utilities, rendered folk psychology, with its beliefs and desires, obsolete. A credence-eliminativist—of whom there may be no clear examples, though Harman (1986) might qualify—will say similar things about the superiority of folk psychology over decision theory, albeit without recourse to the language of obsolescence; presumably they'll think decision theoretic models were stillborn from the start rather than eventually outmoded.

What about views on which both beliefs and credences exist, but which differ on their relative fundamentality? Corresponding to the view that credences are more fundamental than beliefs would be the view that while there are purposes for which it's fruitful to appeal to folk psychological models in prediction and/or explanation—maybe folk psychological models are, for many purposes, more tractable than decision theoretic ones, while still being accurate enough—whatever success they have can be explained by subsuming them within decision theoretic models. On this view, the relationship between folk psychological models and decision theoretic ones is analogous to the relationship between mechanical models that leave out friction and air resistance, and those that include it. While there are many purposes for which we do best to ignore friction and air resistance, we can offer a deeper explanation of *why* this strategy of idealization works by using models that *do* include friction and air resistance, so that we can show when their contribution to observable parameters of interest—e.g., how fast an object will fall—is small. A "Lockean" about belief—someone who thinks belief just is credence above a certain threshold—who thinks it's often most illuminating to explain people's actions by reference to their beliefs, rather than their credences, is naturally interpreted along these lines.[2] Moreover, decision theoretic models have broader scope than folk psychological models; there are phenomena we can explain using decision theory that we simply can't explain using folk psychology (just as the motion of tapioca balls through tea can't be explained without appeal to resistance).

It's not too hard to see how views on which beliefs are more fundamental than credences would tell the opposite story. On such views, there are some aspects of human behavior—what we do in casinos, perhaps—that are more illuminatingly explained using decision theory than folk psychology. But whenever a decision theoretic model explains why someone does something, there's a clunkier folk psychological model that explains the success of the decision theoretic one. Maybe the gambler has a body of beliefs about probabilities,[3] as well as beliefs about how desirable various amounts of money are, and we could explain when he holds and when he folds by reference to a folk psychological model that includes such beliefs, though the explanation is clunkier than the one that skips straight to modeling him directly in decision theoretic terms. Moreover, folk psychological models have broader scope than decision theoretic ones—there are phenomena we can explain using folk psychology that we simply can't explain using decision theory.

What about "dualist" views, according to which neither beliefs nor credences are more fundamental than the other? In the language of modeling, there are two interestingly different scenarios which could naturally be described in these terms. The first, which I take it most dualists in the literature have in mind,[4] is a scenario where there are some phenomena best explained by decision theoretic models, and other phenomena best explained using folk

[2] While they don't present their view this way, this seems to me a helpful lens through which to interpret Fantl and McGrath (2009). On their view, whenever you can explain why somebody (rationally) did something by appeal to their beliefs and desires, there's a more fundamental explanation in terms of credences and utilities, and which can be used to show why even slight changes in their credences would have led to the same action. See also Christensen (2004), a Lockean who's explicit that while it's often most convenient to talk about what people believe, explanations of their behavior and mental life in terms of credence are more fundamental.

[3] This is how Moon and Jackson (2020) explain credence as a species of belief.

[4] E.g., Buchak (2013a), Jackson (2019), Weisberg (2020).

psychological models, and in neither case can these explanations be subsumed within the other framework. But it's also conceivable that decision theory and folk psychology could end up explaining the *same* phenomena. In many areas of philosophy it's standard to discuss the possibility of two putatively distinct theories or frameworks being "notational variants"; differing only superficially, and not substantively.[5] If it turned out that for any arbitrary folk psychological model, we could build a decision theoretic model that would do the same explanatory work, *and* vice versa, then folk psychology and decision theory would plausibly be notational variants.

A central reason why it's very hard to compare the explanatory power of decision theory and folk psychology—and why I won't come to any firm conclusions in this chapter about either framework definitely being able to do the work of the other, or not—is that it's not clear just how to draw the boundaries of either framework. Because folk psychology is informal, it's unsurprising that its boundaries are vague. Does it count as subsuming a decision theoretic model M within folk psychology if we say: "the agent *believes* that M is the appropriate model for her circumstances, and *desires* to act in whatever way is expected utility maximizing according to M"? Or is that somehow cheating? This strikes me as a tricky problem, and adjudicating questions like it will almost certainly call for some degree of stipulation.

But even when it comes to decision theory, the borders can be blurry. While there are paradigmatic examples of explanations that fall into the Bayesian, decision theoretic camp—cases where some change in belief is explained by reference to a model in which the change is the result of conditionalization, or cases in which some intentional action is explained by reference to a model in which that action maximizes expected utility—there are also cases where it's tempting to appeal to degrees of belief in explaining some psychological phenomenon, and yet we lack the mathematical formalism of conditionalization or expected utility theory to undergird the explanation. David Christensen (2004, p. 131) provides some nice examples:

> But beliefs are also universally invoked in explanations of psychological states other than beliefs (and other than preferences). We attribute our friend's sadness to her low confidence in getting the job she's applied for. We explain a movie character's increasing levels of fear on the basis of his increasing levels of confidence that there is a stranger walking around in his house.

[5] E.g., in philosophy of physics (North (2009)), linguistics (Johnson (2015)), and metaphysics (Whittle (2020)).

> The connections between beliefs and other psychological states invoked in such explanations are, I think, as basic, universal, and obvious as the central connections between beliefs and preferences that help explain behavior.

In the book from which this quote is drawn, Christensen explicitly argues that degrees of belief are more fundamental than beliefs. But he is *not* naturally interpreted as saying that anything you can explain with folk psychology, you can explain better with formal models of expected utility maximiziation; becoming increasingly sad or frightened doesn't seem like a choice—optimal or not—and it's not the kind of phenomenon normally thought of as within the purview of decision theory. So however we classify the modeling framework of which credences are a part, it should allow for the possibility of models that posit connections between degrees of belief and emotional states; perhaps "decision theory" is too narrow a label.

All that was just stage setting; I've been laying out schematic victory conditions for various views about the relative fundamentality of belief and credence. I haven't yet said anything about substantive reasons to accept or reject any such view. So it's to such substantive questions that we now turn. My aim in this chapter will be largely negative—I'll try to show that popular arguments that purport to tell against credence-first (or credence-only) views are too quick. I'll first consider an argument that has to do with the complexity of reasoning in a Bayesian manner, and will then consider an argument to the effect that there are normative phenomena that require folk psychological explanations rather than decision theoretic ones.

5.2 Foreclosing Possibilities

Many writers have thought that if we want models not just of human behavior, but of human reasoning, then the Bayesian framework isn't up to the task. Why not? Because it's too complicated for limited agents like us to reason in a way that mirrors the structure of Bayesian models. Here are a few representative quotes:

> One can use conditionalization to get a new probability for P only if one has already assigned a prior probability not only to E but to P & E. If one is to be prepared for various possible conditionalizations, then for every proposition P one wants to update, one must already have assigned probabilities to various conjunctions of P together with one or more of the possible evidence

> propositions and/or their denials. Unhappily, this leads to a combinatorial explosion, since the number of such conjunctions is an exponential function of the number of possibly relevant evidence propositions. In other words, to be prepared for coming to accept or reject any of ten evidence propositions, one would have to record probabilities of over a thousand such conjunctions for each proposition one is interested in updating. To be prepared for twenty evidence propositions, one must record a million probabilities. . . . In the rest of this book I assume that, as far as the principles of revision we follow are concerned, belief is an all-or-nothing matter. I assume that this is so because it is too complicated for mere finite beings to make extensive use of probabilities. (Harman, 1986, pp. 26–7)

Richard Holton defends a position along much the same lines as Harman's:

> A creature that had credences would benefit from the ability to keep many options open at once: such a creature would be able, for instance, to keep in play four mutually incompatible outcomes, regarding each with a credence of 0.25. But such abilities come at a price. . . . Unless their powers of memory and reasoning are very great, those who employ credences risk being overwhelmed by the huge mass of uncertainty that the approach generates. (Holton, 2014, p. 2)

While Harman and Holton both hold that degrees of belief are either nonexistent or reducible to outright beliefs, other writers appeal to similar considerations in defense, not of a belief-first view, but of dualism:

> If we had infinite cognitive resources, then we'd have no need for an attitude of outright belief by which to guide our actions, for we could reason in an ideal Bayesian manner on the basis of our credences and preferences alone. But such reasoning isn't feasible for cognitively limited agents like us, and so we need an attitude of outright belief or of settling on the truth of propositions, so as to limit what we consider in our reasoning to possibilities consistent with what we have settled on. (Ross and Schroeder, 2014, pp. 30–1)

I'll argue that the line of thought expressed by the writers above is roughly half right. The kinds of considerations they raise really do tell against a very ambitious version of the Bayesian picture. On this view, for each agent, there's a single, vast Bayesian model—a single, vast state space Ω, a single, vast algebra $\mathcal{F}$ with its associated probability function P, along with a single, vast set of options and a utility function necessary for calculating expected utilities—that could in principle be used to explain not only the agent's outward behavior but also the inward reasoning process by which they arrive at that behavior, in any arbitrary situation the agent might find themself in. While I'm agnostic about whether anyone would explicitly endorse this view, it doesn't strike me as a straw man; in particular, I think Christensen (2004) invites this interpretation, as does Schwarz (2018, 2021). But for my own part, I'm happy to grant that there must be models of our reasoning processes that capture the fact that they're less computationally demanding than this ambitious Bayesian picture would have it. But where it seems to me that each of the above writers go wrong is in identifying the Bayesian framework for modeling human cognition with this very ambitious version of it. A more modest Bayesian can suspect that decision theory is a strict improvement over folk psychology, without having to think of us as performing impossibly demanding computational tasks in deciding what to have for breakfast.

I've already hinted at a big part of the strategy with my repeated use of the word "vast." Rather than thinking of agents as always solving computationally demanding "grand-world" choices—decision problems in which the outcome space always includes all conceivable ways the rest of world history might progress, depending on all conceivable ways the agent might move their body at the time of choice—the modest Bayesian can think of them as solving much easier "small-world" choices—e.g., which of various breakfast cereals to eat, whether to call or fold in some game of poker. The Bayesian modeler can *agree* with the writers just quoted that limited agents like us need to make a wide range of non-trivial assumptions—about what our options are, what outcomes they might lead to, and how good or bad those outcomes would be—to tractably reason about what to do. She can just insist that those assumptions are properly reflected in the sorts of small-world Bayesian models we can actually write down and solve. Folk psychological notions of belief—understood as something over-and-above or distinct from the fact that small-world models have non-trivial state spaces Ω, in which plenty of logical possibilities are not represented—needn't come into the picture at all.

Of course, a Bayesian modeler who makes this move will have to concede that there must be deeper explanations of *how* we manage to settle on particular small-world representations of our decision problems. This might seem like a substantial retreat. Isn't all the interesting work, then, to be done in explaining how we manage to identify some small-world problem as the one to solve? Well, not *all*. Divide and conquer is a fine explanatory strategy. A modest Bayesian can treat the process by which we settle on small-world decision problems as a kind of black box, and then use her decision theoretic framework to explain how, given the output of that black-box process, we ultimately decide what to do.[6] Such a modest Bayesian could reasonably think of herself as doing real explanatory work, even while conceding that it would be very nice to have some understanding of the workings of the black box.[7]

Now, if we already *had* some alternative framework for modeling the process by which we settle on small-world decision problems, *and* that alternative framework built models using folk psychological categories like belief and desire, then I think it's fair to say that the resulting picture would vindicate a broadly "dualist" picture about folk psychology and decision theory. But it seems to me that's not our situation. My read of the literature is that while there are a variety of approaches to modeling how agents think about small-world decision problems, including alternatives to orthodox Bayesian decision theory, *everybody* has to make substantial appeal to black-box processes when it comes to the question of *which* small-world problem an agent faces; it's not as if we can tell some neat story about that process if only we avail ourselves of folk-psychological tools.

For example, consider prospect theory, famously proposed by Kahneman and Tversky (1979). Prospect theory can smoothly account for some phenomena that are puzzling for traditional decision theory, e.g., the fact that agents sometimes seem to be risk-averse in gains, but risk-*seeking* in losses. Prospect theoretic models represent agents as treating some outcome as the "status quo" or "default"—the outcome relative to which better outcomes count as gains, and worse ones count as losses. The utility functions of prospect theoretic agents have an inflection point at the default—above the default, the marginal utility (of an additional dollar, for example) decreases, as in

[6] See, e.g., Norby (2015), who defends a view along these lines. He suggests that we shouldn't think of degrees of belief as stored, stable features of an agent, but instead as constructed on the fly for use in particular choice situations.

[7] To be fair, even for small-world problems, *exact* Bayesian inference is extremely computationally demanding. As Cooper (1990) proved, computing the exact posterior probabilities in a Bayesian belief-network is NP-hard. Imagine you're playing poker against two people, and you have some prior probability distribution over their possible hands and behaviors. If the way you decide what to do is by computing the exact posterior probability that they have various cards, given their behavior (calling, betting, etc.), then as you double the size of the problem (make it a six-person game instead of a three-person game) the computational resources needed to solve it increase by much more than a factor of two (I'm assuming here that $P \neq NP$). That suggests that however *we* decide what to do, it's more efficient than that. But there is a rich literature on approximate Bayesian inference, with the upshot that it *is* possible, given a prior and some evidence, to efficiently come up with a posterior that, while not exactly right, differs from the exact answer only by a small amount. See Alquier (2020) for a recent overview. Given the perspective urged in this book, it strikes me as not much of a retreat at all if the Bayesian modeler must think that Bayesian models of cognition aren't *exactly* right, and that in the best case scenario for such models, a deeper neuroscientific understanding would reveal that the brain is actually computing some algorithm that *approximates* Bayesian inference.

traditional economic models of risk aversion. But below the default, marginal utility is increasing; if the default is \$0, then going from \$−5 to \$−4 involves a smaller increase in utility than going from \$−4 to \$−3. For my purposes, the crucial point is that the process by which agents come to see an outcome as the default is not *itself* captured by prospect theoretic models. While it's perhaps not *quite* fair to call it a black box—we can make *some* generalizations about how agents' perceptions of defaults can be manipulated—it's certainly true that in practice prospect theory involves a kind of divide and conquer strategy; the precise formal explanations of choice all come into the picture only after defaults have been fixed, and the process by which the default is fixed is one about which only vaguer, more qualitative claims can be made.[8]

In short, only the ambitious Bayesian modeler—the one who's not willing to settle for a strategy of divide and conquer—faces pressing problems of complexity. While the modest Bayesian modeler leaves extremely important questions unanswered, those questions aren't any *better* answered by folk psychology. So the truism that we're computationally limited doesn't clearly put any pressure on the idea that decision theoretic conceptions of belief are a strict advance over their folk psychological ancestors.

5.3 When High Probability Is Not Enough

A very different kind of argument for the necessity of a non-decision-theoretic conception of belief has been given by Lara Buchak (2013a), who draws on a line of scholarship on statistical evidence in legal theory. I'll first describe it in abstract terms, before offering some of the examples to which it is meant to apply. The argument runs something like this:

1. There are pairs of cases involving evidence of guilt or liability that are formally indistinguishable from the decision theoretic perspective; they

[8] See also, e.g., Johnson et al. (2022). They propose "Conviction Narrative Theory" as an alternative to decision theory as a descriptive theory of how agents make choices under radical uncertainty. While it's a rich paper that's impossible to do justice to in a footnote, the basic idea is that agents reason about what to do using narratives, where those narratives generally don't involve assigning probabilities. But it seems to me that "narratives" in their theory play a very similar role to small-world problems in the Bayesian framework. Once you've got a small set of candidate narratives, you can evaluate which best explains your data, and then act in ways that make sense given the winning narrative. But by the same token, once you've got a small-world decision problem, you can figure out which action maximizes expected utility. And just as Bayesian decision theorists generally don't have all that much to say about how we identify which small-world problem to solve, Johnson et al. don't have much to say about how we identify which narratives to evaluate. But if they just treat the process by which we identify narratives as candidates for evaluation as a black box, then it's not clear to me that their approach has any explanatory advantage over that of the modest Bayesian who treats the process by which we settle on small-world problems as a black box.

would be modeled identically by a decision theorist as cases where some evidence makes it highly probable that someone is guilty of some offense.
2. But the pairs differ with respect to crucial normative features—in one member of the pair, the evidence justifies blaming or punishing the person who's highly likely to be guilty, while in the other member, it doesn't.
3. This difference is best explained by an alternative, non-decision-theoretic framework for thinking about how evidence can justify belief and/or action.

While I'm not convinced there are any pairs of cases for which we should accept all of 1–3, I'll suggest that different putative such pairs demand different responses. For some pairs, we should accept that they differ with respect to whether blaming or punishing is appropriate, while denying that they really are identical from a decision theoretic perspective. That is, we should hold that decision theory is up to the task of vindicating the genuine normative distinction between the cases. For others, however, I'll suggest that we should be looking for an error theory, rather than a vindication of our normative judgments.

To give a feel for where I'll go, I think the debate over statistical evidence is in some ways analogous to Thomson's famous trolley problem (1985). While most people intuit a distinction between Thomson's two cases—diverting a trolley from a path in which it will kill five to a path it will kill only one seems OK, but pushing someone into the path of a trolley to stop it from killing five seems wrong—attempts to provide a theoretical vindication of that distinction have generally been unpersuasive, involving baroque, inelegant proposals that turn out to fail to vindicate our judgments in some further elaboration of the case. Moreover, as proposals become more baroque to capture the putative data, they become correspondingly less plausible. The idea that there's a strong prima facie duty against killling—stronger than the duty against letting die—has some intuitive appeal. So does the consequentialist idea that deaths are equally bad and equally worth preventing whether they are the result of action or inaction. But the normative principles one must accept in order to vindicate the distinction between Thomson's two cases, as well as the various elaborations thereof, have considerably less independent appeal.[9] Thomson

[9] For instance, the "Doctrine of Triple Effect," defended by Kamm (2007):

> A greater good that we cause and whose expected existence is a condition of our action, but which we do not necessarily intend, may justify a lesser evil* that we must not intend but may have as a condition of action. (p. 118)

herself ended up rejecting the idea that her two cases had importantly different normative structure (2008) and of course consequentialists were committed to rejecting that idea from the start. Likewise, when it comes to statistical evidence, when we see just how poor are the prospects for an elegant, productive theoretical framework which could illuminate the difference between "mere statistical evidence" and whatever is adequate to justify blame and/or punishment, we should be open to the idea that the category of "statistical evidence" doesn't come close to carving nature at its normative joints. Mere statistical evidence may often be inadequate to justify punishment or blame, but when that's so, it's so for relatively superficial reasons that don't admit of some unified theoretical treatment.

As all that was rather abstract, it will help to have some of the pairs of cases in hand.

BUSES—STATISTICAL: A bus is driving erratically, and hits and kills a woman's dog. While the woman was unable to identify any distinguishing features of the bus, 80% of the buses in the area are owned and operated by the Blue Bus Company. She sues the Blue Bus Company, arguing that there's an 80% chance they're at fault.

BUSES—EYEWITNESS: A bus is driving erratically, and hits and kills a woman's dog. The woman was unable to identify any distinguishing features of the bus. The Blue Bus Company and the Grey Bus Company each own and operate 50% of the buses in the area. But on this particular road, all vehicles with more than two axles (such as buses) must enter a weigh station. On the day of the accident, only one bus was weighed, and the weigh station attendant logged it as a blue bus. The defense attorney for the Blue Bus Company then provided compelling evidence that in the past, the weigh station attendant—having poor vision—has only logged the color of the bus correctly 80% of the time.

The original Blue Bus case is from Tribe (1971), but the contrast with the eyewitness version of the case is from Wells (1992), who presented versions of both cases—as well as some others—to eighty psychology undergraduates, and found that while in both versions of the case they overwhelmingly reported an 80% subjective probability that the Blue Bus Company was at fault, they were much less likely to recommend a finding of liability in the original case than in the eyewitness version of it.[10]

[10] 8.2% of subjects recommended liability in the original case, while 67.1% of subjects recommended liability in the eyewitness version (p. 742).

Here's another pair of cases:

PRISONERS—STATISTICAL: One hundred prisoners are exercising in the prison yard. Ninety-nine of them attack the guard, putting into action a plan that the one-hundredth prisoner knew nothing about. The one-hundredth prisoner played no role in the assault and could have done nothing to stop it. There is no further information that we can use to settle the question of any particular prisoner's involvement.[11]

PRISONERS—DIRECT EVIDENCE: A single prisoner attacks a guard, and while the guard is taken unawares, the attack is observed by a third party. On the basis of the observation, it's 99% likely that Jones was the attacker. (Does it matter how this observation is made and analyzed? Does it matter whether a highly reliable but fallible eyewitness consults her memory, or highly reliable but fallible facial recognition technology analyzes video footage?) There is no further information that we can use to settle the question of any particular prisoner's involvement.

While I don't know of psychological research specifically studying this pair of cases, the stylized fact assumed by much of the literature is that people are much more comfortable finding the prisoner guilty beyond a reasonable doubt in the second case than in the first.

These contrasts are, prima facie, tricky to explain in a decision theoretic framework. The most obvious, flatfooted way of thinking about burdens of proof in a decision theoretic framework is as probabilistic thresholds, so that evidence will warrant a finding of liability when, in light of that evidence, the probability that the accused is at fault is above the crucial threshold (usually understood as 50% for civil liability, and something much higher for criminal liability).[12] So if there are pairs of cases that are otherwise similar—in particular, if they are similar with respect to the probabilities of guilt—but differ with respect to whether the evidence warrants a finding of liability, then it's tempting to look for some alternative framework for thinking about evidential support—one that would explain why statistical evidence often fails to support *belief* in guilt or *knowledge* of guilt, even when it does support high probability of guilt—to capture the difference.

To be clear, that's not the *only* way to vindicate the contrasts. In fact, much of the literature has proposed explanations with a much less

[11] This case goes back to Cohen (1977), though I've taken the wording from Ross (2021).

[12] See Hedden and Colyvan (2019) for an explicit defense of this often implicitly assumed picture.

epistemological flavor. E.g., Richard Posner (1972) argues that letting the plaintiff prevail in cases like Blue Bus would create perverse incentives for companies with less market share than their competitors, and would inefficiently burden larger companies. Maybe so. But if *this* is what our judgments are tracking, then they don't motivate any revisions to our thinking about evidential support. Along similar lines and more recently, Enoch et al. (2012) offer an incentive-based explanation of why it (often) makes sense to discount statistical evidence. Very roughly, they argue that individuals have less control over what merely statistical evidence applies to them, and so have less incentive to modify their behavior in response to the threat of being held liable on the basis of such evidence. Without taking a stand on how successful this explanation is, it too is epistemologically non-revisionary. A helpful analogy might be to the famous "exclusionary rule" in American constitutional law.[13] The rule excludes illegally obtained evidence from being used against a defendant. While the rule has been much debated, one common defense involves appealing to the incentives it creates; police officers who know that illegally obtained evidence won't be used in court have reason to avoid obtaining evidence illegally, if they want their arrests to stick. This sort of justification of the rule is very different from an epistemological one, which would look—unpromisingly, I hope you'll agree—for some sense of evidential support in which illegally obtained evidence doesn't *really* provide evidential support for guilt. Writers who try to vindicate the distinction between pairs like Blue Bus—Statistical and Blue Bus—Eyewitness by appeal to the long-term consequences—especially with regards to incentives—of using this or that rule needn't endorse any novel epistemological theses; in effect, they are denying that the cases really *are* identical from a decision theoretic perspective. Once you factor in *all* the utilities—including the long-run utilities from adopting this or that rule—we can distinguish cases that initially look similar from a decision theoretic point of view.

But suppose we're unconvinced by these sorts of responses, and we're in the market for a more principled, theoretical explanation of the contrast between the paired cases. Martin Smith (2018) defends an interpretation of legal standards of proof that makes use of a non-probabilistic notion of *normalcy*. While his paper (and 2016 book) is rich and I won't be able to do it justice here, the basic idea is that among improbable outcomes, we can distinguish between normal outcomes that don't call out for any special explanation, and abnormal ones that do. It's unlikely that you'll win the lottery, but there's

[13] See Calabresi (2003) for some discussion.

nothing abnormal about you turning out to be the winner; if you do turn out to win the lottery, there won't be some special explanation about what distinguished you from all the losers. By contrast, when a young child develops osteoperosis—a disease normally only found among the elderly—some special explanation is called for, even if the probability of a young child developing osteoperosis is no lower than the probability of an arbirarily selected lottery player winning. The application to legal burdens of proof is that to meet a burden of proof, it should not just be improbable for the evidence to be as it is and yet the accused to nevertheless be innocent—it should be *abnormal*. Smith calls the relation that evidence bears to a claim when it renders the falsity of that claim not just improbable, but abnormal, "normic support."

Smith's view can, I think, be treated as a representative example of the strategy of appealing to folk-psychological, non-decision-theoretic notions of belief and knowledge to make sense of blame and punishment. Smith holds that justified belief and knowledge require normic support; in effect, he aims to provide a theoretical vindication of the very verdicts about cases that Buchak (2013b) appeals to in arguing for the inability of decision theory to account for facts about justified punishment. And while many other writers share Smith's and Buchak's judgments about cases, I know of no other treatment of these matters as systematic and general as Smith's.

As applied to the paired cases, Smith's thought is that in BLUE BUS—STATISTICAL, there would be nothing abnormal about the bus being one of the 20% of grey buses that operate in the area, while in BLUE BUS—EYEWITNESS, were the eyewitness to get the color wrong—even if that happens 20% of the time—there *is* something abnormal about that. There will be some explanation of how his normal color vision process went wrong—perhaps something about the nature of the background lighting—even if that explanation is one that's frequently applicable. And likewise in the two versions of PRISONERS. Basically, whenever there is merely statistical evidence in favor of someone's guilt, that evidence renders their innocence merely improbable, but not abnormal.

I think Smith is absolutely onto something, and the distinction between abnormal and merely improbable scenarios strikes me as a highly fruitful one.[14] In particular, I think the idea that we often ignore "abnormal" possibilities by default can explain a lot of otherwise puzzling facts about which claims we're willing to assert outright, and which we're only willing to assert to be probable, e.g., that we're hesitant to assert outright that this ticket will lose

[14] In addition to Smith, see Goodman and Salow (Forthcoming), who introduce an ingenious framework for modeling inductive knowledge in which normality plays the central role.

the lottery, but more willing to assert outright that an arbitrary child won't get osteoperosis.[15]

But why should we think normic support, in Smith's sense, has anything to do with the justification of action in general, or punishment in particular? Prima facie, it's hard to see why it should. To design a justice system in which burdens of proof function as Smith recommends is, in effect, to hold that judicial mistakes—in particular, cases in which the innocent are held liable—are less regrettable and less worth preventing when they have special explanations than when they are the result of pure chance. But it's hard to see why this should be. Imagine all one hundred people are convicted, including the one innocent person, in PRISONERS—STATISTICAL. And now imagine that PRISONERS—DIRECT EVIDENCE is repeated one hundred times, in each case leading to a conviction, and—unsurprisingly, given the large number of repititions—in one case leading to a false conviction. Are these two convictions of innocent people all that different, from an evaluative point of view? Is the latter case any less bad than the former? To think that a legal system should contain burdens of proof that are sensitive to facts about normic support above and beyond facts about probabilistic support is to think we have more reason to minimize the latter sort of mistake than the former.[16] But by my lights, this would be a legal distinction without an evaluative difference.

This judgment is reinforced by reflection on how odd it would be to directly care about the abnormality/improbability distinction when it comes to actions other than punishment. Imagine two diseases, *Inexplicosis* and *Comprehensoma*, both of which cause identical and generally fatal symptoms to those who suffer from them, and both of which are identically uncommon in the population (let's say that in each case, only 0.0001% of people suffer from them). The difference is that if you get Inexplicosis, we won't be able to identify any deeper explanation of why you got it while most other people didn't, whereas with Comprehensoma, autopsies standardly reveal anatomical irregularities that, while only diagnosable post-mortem, are generally agreed to cause the condition. That said, it's no less bad when somebody gets Comprehensoma than when somebody gets Inexplicosis. We should be just as welcoming of a cure for either condition, and a marginal reduction of *n* cases

[15] See also Phillips and Knobe (2018), who don't use the language of "normality," but are naturally read as offering evidence about the psychological underpinnings of the concept Smith is appealing to.

[16] Here I'm making a similar argument to that made by Enoch et al. (2012), to the effect that the role of epistemological categories like knowledge in the legal system should be merely instrumental; to the extent that the law cares about knowledge or evidence, it should be only as a way better achieve tradeoffs between goals like avoiding punishment of the innocent and achieving punishment of the guilty.

of Comprehensoma is no more or less valuable than a marginal reduction of *n* cases of Inexplicosis. According to Smith's theoretical framework, someone's contracting Inexplicosis is merely improbable, while someone's contracting Comprehensoma is abnormal. But I hope it's plausible that this distinction isn't one that we should (directly) care about when it comes to practical questions about, e.g., how to allocate research funding, or how to structure our health system.[17] By a similar token, it seems to me that the distinction between abnormal cases in which innocents are explicably held liable, and merely improbable cases in which they are inexplicably held liable, isn't one that we should (directly) care about in thinking about how to structure a legal system.

Suppose a framework like Smith's—one in which values interact not with probabilities, but with levels of normic support—were able to do a good job of unifying our judgments about justifiable punishment and blame, while being inapplicable to other sorts of action (e.g., how to direct health funding). In this case, I submit that we'd face a tradeoff that is familiar in other areas of philosophy, between a more "particularist" approach that better fits our judgments about particular cases but which looks somewhat arbitrary and ad hoc when stated at the level of general principle, and a "methodist" approach that has more elegance and theoretical unity, at the cost of running roughshod over many of our particular case judgments.[18] Here the methodist would insist on using decision theory no matter the context, while the particularist would be happy to think about normic support when it comes to justifying punishment, but probabilistic support when it comes to other decisions.[19] Depending on one's philosophical temperament, one might reasonably find oneself pulled in either direction. But as it turns out, I think the supposition with which I began this paragraph is false; even just focusing on our judgments about appropriate

[17] Why do I hedge by including the word "directly"? Because it might be that explicable conditions are easier to treat or cure than inexplicable ones. That would be a fine reason to direct more resources to prevention/treatment of Comprehensoma.

[18] I borrow the language of "particularism" and "methodism" from Chisholm (1973). Unlike Chisholm, I think the distinction between particularism and methodism is better thought of as a matter of degree than as a neat binary.

[19] This is, roughly, the view defended by Buchak (2013a), though she doesn't endorse any particular theoretical framework, such as Smith's, for thinking about what it takes for evidence to justify blame and/or punishment. Rather, she just endorses the idea that the role of evidence in the justification of blame and punishment is systematically different from the role of evidence in the justification of other sorts of action. It's also similar to the view defended by Moss (2018), who holds that some decisions are properly evaluated using decision theoretic norms, while others—such as, but not limited to, decisions about punishment—are properly evaluated by appeal to norms that make reference to knowledge. I criticize this aspect of Moss' view in my Greco (2020), along lines similar to my criticism of Smith's view here; the view can't give a satisfying independent motivation for why different theoretical frameworks are appropriate in the different contexts.

blame and punishment, no neat generalizations about statistical evidence or normic support have all that much unifying power.

To see why this is so, we'll need a few more examples on the table. In fact, the American legal system *has* confronted cases very much like Blue Bus, and while the body of case law is vast and hard to summarize, I think it's fair to say that no neat generalizations about categorical differences between "mere statistical evidence" and direct evidence emerges from it.[20] For instance, there have been many civil cases involving asbestos exposure. As the harms of asbestos inhalation generally don't accrue until decades after the exposure, there's usually a paucity of direct evidence about the nature of the exposure. And in some such cases, courts have allowed plaintiffs to recover compensation by appeal to fact patterns that are highly reminiscent of the Blue Bus case. Here's a recent Delaware court (reaching a similar judgment) describing an earlier line of asbestos cases:

> In one Fel-Pro gasket case, the court denied summary judgment when the plaintiff showed through circumstantial evidence that during the period of the plaintiff's alleged exposure, 98% of Fel-Pro's gaskets contained asbestos. The court found that this overwhelming probability of asbestos exposure was enough to link the plaintiff's exposure to asbestos-containing gaskets. None of the Fel-Pro gasket cases foundered on the shoals of failing to show direct evidence of exposure.[21]

In the case under discussion the plaintiff could prove that they had extended exposure to Fel-Pro gaskets, but couldn't provide direct evidence that there was asbestos in the gaskets; not all of Fel-Pro's products at the time contained asbestos. But the statistical evidence that the vast majority of Fel-Pro gaskets sold at the time did contain asbestos was considered adequate to shift the burden onto Fel-Pro to show that the particular gaskets the plaintiff worked with *didn't* contain asbestos. For my own part, this doesn't strike me as a miscarriage of justice—not by a long shot. But I struggle to find a deep, principled distinction between this sort of case and the original Blue Bus case.

To take another class of case, Ross (2021) persuasively argues that DNA evidence, as used by American courts, is a counterexample to the thought that merely statistical evidence of guilt is generally insufficient to establish criminal liability. Ross discusses cases of "cold-hit" DNA evidence, where DNA

[20] See Ross (2021) for a defense of this claim.
[21] See Droz v. Hennessy Indus., No. 211 (Del. Mar. 28, 2022).

is collected at the scene of a crime, and then is found to be a match (with extremely high probability) to DNA already contained in some database. In such cases, it generally hasn't been held that additional, not-merely-statistical evidence is required to establish criminal liability. While DNA evidence is certainly rebuttable—Ross mentions a case where a defendant successfully raised the possibility that the DNA might have been his brother's—in the absence of such rebuttals, convictions on DNA evidence alone are not unusual. Now the probabilities in cases involving DNA evidence are generally much more extreme than the 0.99 probability of guilt in PRISONERS. Still, writers who take cases like PRISONERS very seriously tend to say that a finding of criminal liability would remain inappropriate even if the probabilities were increased arbitrarily, so this seems like an inadequate basis on which to distinguish cold-hit DNA cases from PRISONERS, for those who want to explain why it would be unjust to convict in the latter case by appeal to facts about statistical evidence.[22]

My aim in bringing up cases of asbestos exposure and cold-hit DNA evidence is to show how difficult a path we'd have to tread to vindicate all the widely held normative judgments about justifiable punishment by appeal to general principles about statistical evidence *as such*. Smith himself concedes that his view is committed to regarding findings of criminal liability in "cold-hit" DNA cases as illegitimate (2018, p. 1214); it is, at least to that extent, revisionary. He doesn't discuss asbestos cases, but I imagine that he'd say much the same thing about the use of statistical evidence in them. Once such cases are on the table, however, the argumentative landscape is significantly different. We're no longer faced with a choice between an approach that's well-motivated but fits poorly with particular case judgments, and an alternative approach that's less well motivated but fits better with particular case judgments. Rather, fit with particular case judgments now looks like more of a wash, unable to count in favor of the normic support framework over the orthodox decision theoretic one. But the orthodox decision theoretic framework still has its advantages of generality and simplicity. So we should be more willing to simply bite bullets in cases involving statistical evidence, as suggested by Hedden and Colyvan (2019), and/or to look for treatments like those of Posner (1972) and Enoch et al. (2012), which aim to accommodate intuitive judgments about such cases within the confines of orthodox decision theory.

[22] For instance, Lara Buchak writes: "But I submit that we never think it justified to blame an individual on the basis of merely statistical evidence: doing so is not merely bad, it is prohibited" (Buchak, 2013a, p. 303).

The arguments of this chapter may seem in tension with the modest, pluralist methodological stance urged throughout this book. Shouldn't a modest modeler be open to the suggestion that we've tried to bring too wide and diverse a range of phenomena within the ambit of decision theory? Shouldn't she celebrate the proliferation of modeling frameworks that might carve off some terrain that ambitious, imperialistic types—economists, especially—have prematurely claimed for decision theory? Not necessarily. Modesty in modeling shouldn't mean abandoning the idea that power and scope are theoretical virtues, or that all else equal it's better to explain more by appeal to less.

A modest modeler should be *open* to the possibility that some phenomena (e.g., what we do in casinos) are best captured by decision theoretic models, while other phenomena (e.g., who we ought to hold criminally liable) are best captured by some other, very different sorts of models for thinking about how evidence can rationalize choices. But she should still respect some version of Ockham's razor, in which neither models nor modeling frameworks are multiplied beyond necessity. I've argued in this chapter that when it comes to questions about how computationally limited agents handle complexity, and questions about whom we ought to hold criminally and civilly liable, there's less necessity for multiplying modeling frameworks—as well as less explanatory power to be gained by doing so—than one might have thought.

6
Inter-Level Coherence

In Alston's "Level Confusions in Epistemology," he argued that many philosophers wrongly conflate claims at different "epistemic levels." Just as we can ask whether I know that P, we can ask whether I know that I know that P, or whether I know that I know that I know that P, and so on all the way up. Likewise with other epistemological concepts, such as probability, evidence, and rationality. We can ask whether P is probable, but also whether it is probable *that* P is probable, or whether it's rational to believe that P, but also whether it's rational to believe *that* it's rational to believe that P. A theme of much recent epistemology has been that claims at different levels are by and large independent, and that failure to appreciate distinctions between levels is the source of many epistemological errors.[1] Some, however, have self-consciously defended various of the putative conflations, arguing, e.g., that there isn't really a distinction between knowing that P and knowing that one knows that P, or between having justification to believe P and having justification to believe one has justification to believe P.[2] It's common in formal epistemology to work in modeling frameworks that impose strong constraints on the relationships between claims at different epistemic levels. The "KK thesis" and its cousin for belief—if you know(believe) that P, then you know(believe) that you know(believe) that P—are often assumed in epistemic logic.[3]

Economists and computer scientists typically go a step further, using models in which the possible bodies of evidence an agent can receive partition logical space—each of her possible bodies of total new evidence is incompatible with the rest, and she's guaranteed to learn at least one of them. This can seem like an innocent assumption. Suppose we're modeling the beliefs of a roulette player. It's natural to use a state space corresponding to each of the possible outcomes of the spin—each of the numbers 0 through 36—and to say that when she

[1] This is a major theme in work inspired by Williamson (2000).

[2] See McGrew and McGrew (1997) for a direct defense of "level connections in epistemology." I've defended similar views in Greco (2014a) and Greco (2014b). See also part II of Smithies (2019a) for a recent, ambitious defense of a wide range of level-bridging epistemological principles.

[3] E.g., in Hintikka (1962).

Idealization in Epistemology: A Modest Modeling Approach. Daniel Greco, Oxford University Press. © Daniel Greco 2023.
DOI: 10.1093/oso/9780198860556.003.0007

observes the result of the spin, she learns which of those outcomes happens. In this case partitionality is almost irresistible: she's guaranteed to learn one of {0, 1, 2...36}, and whichever she learns, that rules out any of the others being true. But such partitional models validate not only positive introspection (the KK thesis), but negative introspection (the thesis that if you don't know that P, then you know that you don't know that P) as well.[4]

And while there have been vocal dissenters, throughout much of the twentieth century it was common for writers in the Bayesian epistemological tradition to explicitly treat higher-order probabilities as unintelligible or trivial.[5] While an exhaustive survey is well beyond the scope of this book, see the footnote below for examples of influential results in formal epistemology that crucially depend on some or another "level confusion."[6]

"Conflating" distinctions between first- and higher-order belief, evidence, probability, or knowledge, is often treated as an idealization—to ignore such distinctions may be a harmless simplification for certain purposes, but it's nevertheless an error.[7] In this chapter I'll explore a strategy for defending modeling frameworks that make such "conflations" that parallels the strategy I used in Chapter 3 to defend the possible worlds framework for modeling information. In that chapter I argued that for the folk psychological modeling framework to do its work—for it to make sense of how beliefs and desires together rationalize action—an agent's beliefs must obey various coherence

[4] See, e.g., Samuelson (2004) and Fagin et al. (2003) for representative examples of partitional models in economics and computer science, respectively.

[5] See Skyrms (1980), who surveys the largely dismissive tradition in twentieth-century formal epistemology towards higher-order degrees of belief, while himself rejecting that tradition.

[6] Some examples:

1. The standard treatment of the Monty Hall problem, on which the probability of receiving a prize upon switching is $\frac{2}{3}$, compared to just $\frac{1}{3}$ if one sticks with one's initial choice, relies on the assumption that when one learns that the prize is not behind a given door, one learns *that* one has learned this. This is essentially a version of the KK principle, according to which one can't gain knowledge without gaining knowledge that one has gained knowledge. See Bronfman (2014), Greco (2019, pp. 90–1) for discussion.
2. Briggs' (2009) argument that a qualified version of the famous "reflection principle" (van Fraassen, 1984) follows from the probability calculus depends on the assumption that an agent's possible evidence propositions form a partition. But as noted in the text, partitionality requires that evidence obeys both positive and negative introspection principles: if P is part of your evidence, then it's part of your evidence *that* it's part of your evidence, and if P is not part of your evidence, is part of your evidence *that* it's not part of your evidence.
3. Good's (1966) famous argument that cost-free information is always valuable similarly depends on a partitionality assumption. See Salow and Ahmed (2019) for explanation and criticism.
4. In a series of recent papers, Nilanjan Das (2019, 2020, 2022) has argued that a wide range of principles in formal epistemology—in addition to some of the previous, he discusses the idea that rational agents are immune to dutch books—crucially depend on partitionality assumptions, and should be rejected on that account.

[7] See, e.g., Williamson (2000, p. 18), Lederman (2018a, p. 937).

constraints like closure and consistency. Apparent counterexamples to those coherence constraints, I argued, are best construed as cases where no one folk psychological model perfectly fits an agent. Moreover, this imperfect fit shouldn't bother us much, since such messiness is typical when we model the world using highly non-fundamental categories.

In this chapter I'll show how a similar strategy can be used to defend modeling frameworks that "conflate" various level-distinctions. We often run into trouble when we try to combine bodies of belief or knowledge that violate inter-level coherence constraints with additional attractive principles about the links between knowledge, belief, action, and assertion. While the troubles aren't quite as stark as the ones we encountered in Chapter 3—we don't get triviality—we do get odd results that don't match what we would expect or recommend agents to do. But we can nicely account for the cases using a broadly fragmentationist, modest modeling strategy—cases it's tempting to describe as witnesses to level-distinctions are often cases in which for some purposes one model fits best (capturing claims at one level), and for other purposes another fits best (capturing claims at another), but each of these models is one that, taken individually, fails to exhibit level-distinctions.[8]

In §6.1 I'll illustrate some difficulties that arise when we combine bodies of information that violate inter-level coherence constraints with attractive principles linking belief and knowledge with action and assertion. In §6.2 I'll offer a different motivation for not working in modeling frameworks that distinguish first-order and higher-order attitudes; such frameworks typically allow not only for a first-order/second-order distinction, but also a second-order/third-order distinction, a third-order/fourth-order distinction, and so on all the way up. But this infinite hierarchy of distinctions introduces a great deal of complexity into our modeling frameworks, without a corresponding gain in explanatory power. In §6.3 I'll argue that certain phenomena that have been thought to put pressure on introspection principles are much less threatening once we are content to be modest modelers.

[8] Of course, I wouldn't want to pursue such a strategy for *all* level distinctions. To take one notable class, I don't think there are any *inter*personal level distinctions I'd want to "conflate"; it's hard to imagine a good reason to work in a framework that can't draw the distinction between knowing that P and knowing that *someone else* knows that P, or between the latter and knowing that someone else knows that *you* know that P. So it might seem that any responsible discussion of level-distinctions would have to pick some specific one(s) to focus on, rather than talking about the category in general. I don't think that's quite right. My aim in this chapter is to illustrate a general strategy for redescribing level-distinctions that, while not universally applicable, should be on our collective radar as epistemologists. I myself am agnostic about just how widely it can be used.

6.1 Inter-Level Incoherent Attitudes in Speech and Action

Consider the following principles:

KK: If you know that P, then you know that you know that P.

BB: If you believe that P, then you believe that you believe that P.

When we model people using frameworks where the above principles can fail, we run into difficulty if we also accept principles linking knowledge and belief with assertion, such as the following:

KAA: If you know that P, then you're entitled to assert that P.

BSA: If you assert that P, then your assertion is sincere just in case you believe that P.[9]

What's the problem? In short, it's that whenever an agent is entitled to assert that P, she's entitled to assert that she knows it. And whenever an agent can sincerely assert that P, she can sincerely assert that she believes it. Versions of this point have been made by many writers who've drawn attention to the awkwardness of insisting that P, while disavowing knowledge or belief that P.[10] To modify an example I've used before (in Greco 2014b), consider the following dialogue:

ALICE: When did Queen Elizabeth die?
BOB: She died in 1603.
ALICE: How do you know that?
BOB: I didn't say I know it.
ALICE: So you're saying you don't know when Queen Elizabeth died?
BOB: I'm not saying that either. I'm saying she died in 1603. Maybe I know that she died in 1603, maybe I don't. Honestly, I've got no idea. But you didn't ask about what I know, did you? You just asked when she died.
ALICE: Do you at least believe that she died in 1603?
BOB: Maybe? You should ask my analyst; she's great at figuring out what I believe.

[9] The literature on norms of assertion is vast—my hope is that the arguments of this section could be adapted to fit with a wide range of positions in that debate. Whether we center knowledge (Williamson, 2000, ch. 10), reasonable belief (Lackey, 2007), certainty (Beddor, 2020, forthcoming), or something else, we'll run into issues like the ones discussed below.

[10] See, e.g., Shoemaker (1994) and Sosa (2009).

Bob seems to be trolling. But if he knows that Queen Elizabeth died in 1603 without knowing that he knows this, and KAA is true, then he's being a cooperative conversational participant; he's asserting all and only that which he's entitled to concerning the question under discussion. And if he believes that she died in 1603 without believing that he believes this, then he may be completely sincere. Since it's hard to hear Bob's half of the dialogue as cooperative conversational behavior, we face some pressure to deny that it really is possible to know without knowing that you know, or to believe without believing that you believe.

Similar difficulties arise when we ask how inter-level incoherent agents will or ought to think, reason, and plan. Consider the case of Matt, who is afraid of flying. Matt believes that airplane travel is extremely dangerous, but also believes that this belief of his is an irrational phobia, unsupported by his evidence.[11] How should we expect him to behave? If his case is a realistic one, we might expect him to avoid air travel, even at great personal expense and inconvenience, but also, perhaps, to seek therapy aimed at treating his condition. But this isn't what we'd predict if we naively took the beliefs we stipulated him to have—a first-order belief about the danger of air-travel, and a higher-order belief about the rationality of that first-order belief—and predicted that he would perform those actions that would best promote his desires, given those beliefs.

Why not? Assume, realistically, that Matt strongly values his continued survival. In particular, he values his continued survival much more than he values his rationality. In that case, naive instrumental reasoning would suggest Matt would *avoid* therapy, at least if it's likely to be effective. For example, suppose Matt hears than a particular therapist has a 100% success rate at treating aviophobia. Given his current beliefs, seeing the therapist is likely to produce an outcome in which (a) his beliefs are rational, and (b) he gets on a plane, flies, and dies. If he prefers a long irrational life to a short rational one, then instrumental reasoning would predict that he won't pursue therapy.

We can make a similar point by considering a case of implicit sexism, briefly touched on in Chapter 3. Versions of the following example have been discussed a great deal in both the psychological and philosophical literature:[12]

[11] I discuss this example in Greco (2014c).

[12] For a philosophical discussion of a more detailed example very much like the one below, see Schwitzgebel (2010, p. 2). The psychological literature on implicit attitudes is vast, but the collected bibliography here is a good start: https://implicit.harvard.edu/implicit/demo/background/bibliography.html.

> The Implicit Sexist: John is an avowed anti-sexist. In particular, he is prepared to defend vigorously the equality of the sexes in intelligence. Yet, in a variety of contexts, John's behavior and judgments are systematically sexist. Concerning the individual women he knows, John rarely thinks they're as intelligent as the men he knows, even when John has ample evidence of their intelligence. In group discussions, John is systematically less likely to pay attention to and take seriously the contributions of women. On the rare occasions when he does judge a woman to have expressed a novel, interesting idea, he is much more surprised than he would have been if a man had expressed the same idea. Still, John is unaware of these dispositions, and he would deny that he had them if asked.

The implicit sexist is sometimes thought of as inter-level incoherent—he believes men are smarter than women, but doesn't believe that he believes this. But do implicit attitudes, such as implicit sexism, really illustrate a distinction between first-order and higher-order belief? To put the question another way, can we fruitfully model the case above as one in which John believes that P, believes that he doesn't believe that P, and consistently acts in accordance with those beliefs? If John believes that P—that men are smarter than women—then he should act as if men are smarter than women. That is, he should behave in ways that would satisfy his desires if it were true that men are smarter than women. In some respects, he does do this. Plausibly, dismissing women's contributions in group discussions would satisfy a background desire to spend time efficiently, if it really were true that women were unable to offer promising suggestions due to lack of intelligence. But he doesn't consistently act as if men are smarter than women. We may assume that he generally prefers to speak the truth. But, as the example stipulated, if the subject explicitly comes up in conversation, he'll vigorously defend the equality of the sexes in intelligence. This behavior would frustrate his background desire to speak the truth, if men really were smarter than women. While we could recast the previous discussion in decision theoretic terms, it wouldn't change the upshot; it's only for certain purposes—for capturing certain aspects of his behavior—that John is fruitfully modeled as believing that men are smarter than women. For others—e.g., for making sense of his contributions in explicit discussions of the topic—he's more fruitfully modeled as believing that the sexes are equal.

It might be objected that I haven't brought John's higher-order belief into the picture—can we explain why he defends the equality of the sexes by appeal to his higher-order belief that he *believes* they are equal? I assumed his overriding desire when the question of the equality of the sexes comes up was to speak truly about the topic under discussion, from which it plausibly follows that, regardless of his beliefs about his beliefs, if he believes men are smarter than women, then saying that men are smarter than women would better satisfy his desires than denying it.[13] But even absent that unrealistic, simplifying assumption, it's hard to imagine what alternative set of values would be such that someone who behaves like John could be fruitfully modeled as consistently acting so as to promote those values given the first-order belief that men are smarter than women and the higher-order belief that he doesn't (or, shouldn't) believe this.

In Chapter 3 we saw that instrumental reasoning is reduced to triviality when the input beliefs are logically inconsistent. What I take the above discussion to suggest is that when the input beliefs are inter-level incoherent, while we don't get triviality, we do get something in the ballpark; higher-order beliefs about one's own beliefs or evidence seem to fall out of the picture, failing to predict certain actions that, intuitively, they should. To return to the case of Matt, we'd expect his higher-order beliefs about the rationality of his fear of flying to make a difference to, e.g., how likely he is to seek therapy. But we can't capture that natural prediction by simply taking both his first-order and stipulated higher-order beliefs, and then asking what would satisfy his desires (or maximize expected utility) given those beliefs.

So how should we understand attributions of inter-level incoherent bodies of belief? Much in the same way we understand attributions of inconsistent bodies of belief. When we describe someone as believing P, but believing that she doesn't or shouldn't believe this, or as knowing P but not knowing that she knows it, the sorts of predictions about behavior we'd ordinarily take that attribution to support can't be captured with a single folk psychological model in which the agent violates inter-level coherence constraints. Rather, it's only by using different models for different purposes—models which, taken individually, needn't exhibit any inter-level incoherence—that we can generate the intended predictions. E.g., perhaps Matt is happily modeled as believing that flying is dangerous for some purposes (e.g., deciding whether to get on the plane) but not others (e.g., deciding whether or not to see the therapist).

[13] This parallels the more realistic assumption, in the case of Matt, that Matt values survival much more than rationality.

6.2 Higher-Order Level Distinctions

When Alston (1980) introduced the terminlogy of "epistemic levels," he used the following three infinite series to illustrate the idea:

1. P.
   S believes that P.
   S believes that s believes that P.
   . . .
2. P.
   S is justified in believing that P.
   S is justified in believing that S is justified in believing that P.
   . . .
3. P.
   S knows that P.
   S knows that s knows that P.
   . . .

The sorts of putative confusions Alston diagnosed concerned the second and third levels. For example, he accuses Chisholm of sliding between questions about what justifies a subject in believing that P, and questions about what justifies a subject in believing that she knows that P. But while the accusation of first level/second level conflations is familiar in the history of philosophy—arguably verificationism rests on such a mistake—it's a curious fact that we see much less discussion of third level/fourth level conflations, or still higher-order ones. The cases I discussed in the previous section were all putative cases of second level/third level distinctions—knowledge without knowledge of knowledge, belief without belief in belief, and belief without belief in rationality. And while I suggested that, ultimately, cases of irrational phobias and implicit sexism aren't best interpreted as illustrating level distinctions in epistemology, I certainly admit that they're intuitively natural to describe in such terms. But while it's straightforward to formally model epistemic level distinctions of arbitrarily high order—modeling frameworks in which KK fails typically allow for n^{th} order knowledge without $n + 1^{\text{th}}$ order knowledge for any n[14]—it's much harder to come up with cases that are intuitively natural to describe in such terms.

[14] See, e.g., Williamson (2000, 2011, 2014).

Suppose we stipulate that Hansel knows that he knows that thesauruses aren't dinosaurs, but he doesn't know that he knows that he knows this, or that Gretel believes that she ought to believe that she ought to believe that flying is safe, but she doesn't believe that she ought to believe this. Under what circumstances could people be rightly described these ways, and how should we expect such agents to think, act, and speak? Of course, I can imagine someone who is disposed to utter "I know that thesauruses aren't dinosaurs," but also disposed to refrain from uttering "I know that I know that thesauruses aren't dinosaurs." Likewise, I can easily imagine someone disposed to utter "I ought to believe that I ought to believe that flying is safe" but not disposed to utter "I ought to believe that flying is safe." But, speaking purely autobiographically, in neither case do I feel like I can go much beyond that. If I'm told that someone believes flying is dangerous, but believes that this belief is irrational, or that they believe men are smarter than women, but don't believe that they believe this, I can construct rough and ready folk psychological models that will generate a reasonably wide range of behavioral predictions, not just concerning dispositions to utter particular sentences. But when it comes to descriptions of agents like Hansel and Gretel, I find myself at a loss. This should make us suspicious; if our informal folk psychological modeling framework really does allow for the sorts of distinctions Hansel and Gretel are stipulated to exemplify, we should have an easier time making sense of such examples.

Perhaps you'll object that all this shows is that we're bad at dealing with complexity (or at least that *I* am). By analogy, there are myriad examples of hard to parse but nevertheless grammatical sentences that most speakers can't understand without a great deal of effort. For example: "fish fish fish eat eat eat." Perhaps attributions of bodies of belief and knowledge that exhibit higher-order level distinctions—attributions of third- but not fourth-order knowledge, for instance—should be understood similarly. Our competence in theory of mind can in principle make sense of such attributions—in particular, it can combine them with desire attributions to generate behavioral predictions—but when it comes to performance, we often fall flat.[15]

In the remainder of this section I'll argue that not only does this response fail, but it backfires, ultimately strengthening the case for thinking that neither informal folk psychology nor its decision theoretic refinements need make arbitrarily high-order level distinctions.

[15] The performance/competence distinction was introduced by Chomsky (1965).

It's true that higher-order attitude ascriptions are more complex and hard to parse than lower-order ones. But that complexity can't explain the difficulty of thinking through descriptions like those of Hansel and Gretel, because it fails to capture a crucial contrast: the contrast between higher-order *inter*personal attitude ascriptions, which are difficult to parse but uncontroversially intelligible, and higher-order *intra*personal attitude ascriptions, which make much less sense.

In 1976, the Kursaal Flyers had their one and only top-20 hit: "Little Does She Know." The opening line is a wonderful example of higher-order attitude ascription in the wild:

Little does she know that I know that she knows that I know she's two-timing me.

The line takes some effort to parse. It's not describing the simple, familiar case in which A is cheating on B, not realizing that B knows about A's infidelity. Rather, it's describing a case in which A cheats on B, B knows that they're being cheated on, A knows that B knows about the cheating, B knows that A knows that B knows about the cheating, but A doesn't know that B knows that A knows that B knows about the cheating. The song elaborates:

She was sharing her spin dryer with a guy in a tie-dye
When she saw my reflection in the chrome
I knew that she'd seen me 'cause she dropped her bikini
The one that I got her, in Rome.

As is clear as the song progresses, while she observed herself being observed, she didn't realize that the speaker had seen her reaction; as far as she knows, the speaker merely saw her cheating, but didn't realize she'd seen him. The song illustrates a case of a complex, difficult-to-parse, but uncontroversially intelligible higher-order attitude ascription.

The example is not unusual. Similar higher-order attitudes play crucial roles in explaining lots of strategic behavior. If I'm playing poker, I'll be well advised to ponder not just what your hand is, and not just what you think my hand is, but also what you think I think your hand is. And a commentator might explain a player's bets by making such higher-order attitude ascriptions. Higher-order *inter*personal attitude ascriptions are complex, but systematically intelligible; with some effort, we can think of explanations for how somebody might come to know that somebody else knows that some third person knows that the original person believes that so-and-so, and those higher-order attitude ascriptions will have a systematic role to play together with desires/utilities in generating behavioral predictions, both in informal folk psychology as well as its decision and game theoretic refinements.

By contrast, consider the following, merely possible song lyric:

Little does she know that she knows that she knows that she knows she's two-timing me.

This would've been unintelligible, and no song-lyric-sized followup would've helped. We can neither imagine a natural backstory that would account for how she'd have gotten into such a state, nor can we appeal to such a state to explain how she might go on to behave. If they'd opened the song with such a clunker, the Kursaal Flyers would never have made it onto the charts.

How can we explain this marked contrast between higher-order interpersonal attitude ascriptions and higher-order intrapersonal attitude ascriptions?

The strategy I propose for making sense of the contrast involves several ingredients. First, we can treat inter-level coherence principles, like KK, BB, and perhaps negative introspection principles as well, as playing a similar role to principles of consistency or closure for belief—as prerequisites that a body of information must satisfy if it is to systematically combine with values to recommend action. This might seem to be enough all on its own, as it gives us a difference between higher-order intrapersonal attitude ascriptions, and higher-order interpersonal attitude ascriptions; the situation described by the Kursaal Flyer's doesn't violate KK, but the imagined alternative with both occurrences of "I" replaced by "she" would. But if the story stopped there, we wouldn't have an explanation of why *some* inter-level incoherent attitude ascriptions—especially, the ones that draw distinctions between the second and third items in Alston's series—seem perfectly intelligible, and capable of generating behavioral predictions in combination with desires.

The second ingredient is fragmentation/modest modeling. Just as we can make sense of attributions of inconsistent or deductively open bodies of belief by interpreting them as attributing various different consistent bodies of belief, which can fruitfully model various different aspects of an agent's behavior, we can do something similar with attributions of inter-level incoherent bodies of belief. When we hear somebody described as knowing that you need to lean into a turn when riding a bike, but not knowing that they know this, we infer that some aspects of their behavior—e.g., how they position their body while riding a bike—are nicely modeled as goal-directed activity relative to a body of information that includes the fact that you need to lean into your turns, but other aspects of their behavior—e.g., how they talk about bike-riding—are best modeled as goal directed activity relative to a body of information that lacks this fact.

The second ingredient threatens to undermine the first—if we can use fragmentation/modest modeling to interpret ascriptions of inconsistent or inter-level incoherent bodies of information, why doesn't it work when we go a few levels up?

This is where the modesty comes in. There isn't any systematic algorithm for translating an attribution of inconsistent or inter-level incoherent beliefs into a set of attributions of consistent, inter-level coherent beliefs, each operative in different contexts. While we manage it, we do so unsystematically, on a case-by-case basis. When I say that Napoleon believes that all animals on the farm are equal, and that he is first among equals, while I've plausibly attributed inconsistent beliefs, we can have some guesses as to which consistent bodies of belief we expect to be operative in which situations; I've painted the beginnings of an intelligible psychological portrait. But not all attributions of inconsistent beliefs are like that. If I say that Joe believes that there are exactly five platonic solids, and Joe also believes that there are exactly 431 platonic solids, while also believing that "platonic solid" is a meaningless expression, it's much harder to make sense of my belief ascription or to combine it with ascriptions of desires to generate behavioral predictions.

And what goes for inconsistency goes for inter-level incoherence. The modified Kursaal Flyers line—little does she know that she knows that she knows that she knows she's two-timing me—is a lot like the attribution of inconsistent beliefs to Joe. Not only does it involve inter-level incoherence—third-order but not fourth-order intrapersonal knowledge—but moreover there's no salient or natural way to extract from it a set of inter-level coherent bodies of belief, each operative in different situations. By contrast, the original line attributes no incoherence, and so no such translation or extraction is required to make sense of it.

6.2.1 Complexity and Epistemic Iteration

In the previous section I suggested that most discussion in philosophy of "epistemic levels" has focused on the distinction between the second and third of Alston's levels, and very little beyond that. In this section I'll consider an objection to introspection principles (whether positive or negative) that does crucially turn on higher-order level distinctions. Very roughly, the worry is that introspection principles require that we have an infinite hierarchy of increasingly complex beliefs (or items of knowledge) if we have just one justified belief (or item of knowledge). But since it is implausible that we have such an infinite hierarchy of increasingly complex beliefs (or items of knowledge), introspection principles lead to skepticism. Since skepticism is false, this is a *reductio*.

Here are some representative statements of the objection, with wording slightly altered for uniformity:

> Another objection turns on the fact that, for any proposition P which *S* is justified in believing, [JJ] requires not only that *S* is justified in believing P, but also that *S* is justified in believing that *S* is justified in believing P. For by a second application of [JJ], we see that [J^2p] implies [J^3p]. Further, [J^3p] implies [J^4p], and so on. It is unreasonable, however, to assume that epistemic subjects actually have such an infinite hierarchy of beliefs. It is implausible, for example, that I can even grasp the proposition indicated by "[J^4p]." (Greco, 1990, p. 262)
>
> Given [JJ], therefore, one has a justified belief only if one actually has an infinite number of justified beliefs of ever-increasing complexity. But most of us find it exceedingly difficult even to grasp a proposition like [J^5p] or [J^6p] in such a series, much less believe it with justification. Consequently, it's very difficult to see how a supporter of [JJ] could resist the conclusion that none of our beliefs are justified. The very ease with which this skeptical conclusion follows from [JJ] gives us a reason to reject it.
>
> (Bergmann, 2006, p. 15)

Even those sympathetic to introspection principles often qualify them to avoid objections like those above. Jonathan Schaffer (2010, p. 234) appeals to a version of KK, but holds that its application may be limited to those with "infinite capacities."[16] To my knowledge, however, those who object to introspection principles on complexity grounds have never spelled out exactly why it's implausible that we have such an infinite hierarchy of belief or knowledge, or whether and how the "increasing complexity" of this hierarchy presents problems. In this section, I'll argue that not only can the defender of introspection principles resist objections like the ones above, but she can turn

[16] Brueckner (2011) objects that if KK only holds for agents with unlimited capacities, its relevance to agents like us is questionable.

the tables; once we're working with a plausible conception of complexity, we'll see that it's the *opponent* of such principles who runs the risk of attributing unrealistically complex bodies of belief or knowledge to epistemic subjects.

There's one model of how humans might store information that would straightforwardly vindicate the complexity argument against introspection principles. If having infinitely many beliefs required storing an infinite list of distinct mental representations (perhaps sentences in the language of thought)—one for p, one for Kp, one for K^2p, and so on—then having infinitely many beliefs would be beyond our capacities. That the infinite hierarchy required by introspection principles involves "increasing complexity" wouldn't even come into the picture.

But many philosophers (as well as thinkers in other disciplines) deny that storing distinct items of information always requires storing distinct representations corresponding to each item of information. It's a familiar point that on broadly dispositional theories of belief, creatures with finite memory may nevertheless have infinitely many beliefs.[17] Is moving away from a list-storage picture of belief enough to defeat the complexity argument against introspection principles?

Not obviously. There's still the worry that such principles require that we have not just any infinite set of beliefs, but one of "increasing complexity." Bergmann (2006, p. 16) presses this point against a version of JJ that is consistent with broadly dispositional approaches to belief. However, without some idea of how the body of information associated with the infinite hierarchy $\{p, Kp, K^2p\ldots\}$ might be stored, it's hard to say whether or not storing that information would be hard for computationally limited agents like us.

Luckily, there are independently well-motivated ways of thinking about complexity that we can import from computer science, and which may shed light on the question. The *Kolmogorov complexity* of a string S is the length of the shortest program (in some reasonable programming language L) that outputs S on a null input.[18] As the definition makes clear, Kolmogorov complexity is a language-relative notion, but in practice the language-relativity matters little; as long as we exclude "gruelike" programming languages, not much depends on the choice of language.[19]

[17] See, e.g., Dennett (1981a) and Stalnaker (1984, p. 68).

[18] See Li and Vitányi (2008) for an introduction to the topic of Kolmogorov complexity.

[19] In a gruelike programming language, there might be no short program that outputs the string "10," while there might be a very short program that outputs the string "29343789786803789 05056783023658347658340."

Just as we can talk about the Kolmogorov complexity of an individual string, we can also talk about the Kolmogorov complexity of a set Γ of strings. While there are various different equivalent ways of generalizing the notion of Kolmogorov complexity to sets, a convenient one is as follows: The Kolmogorov complexity of a set Γ is the length of the shortest program P that tests for membership in Γ. That is, the shortest program P such that for any string $\gamma \in \Gamma$, P outputs 1 on input γ, while for any string $\delta \notin \Gamma$, P outputs 0 on input δ.[20]

With these materials, we can ask a tractable formal question that is plausibly relevant to whether or not the complexity argument against epistemic iteration principles succeeds: which has a greater Kolmogorov complexity, the infinite set {"p," "Kp," "K^2p," "K^3p"...}, or some finite initial subset of that set, say {"p," "Kp," "K^2p"}? Perhaps surprisingly, the answer is that the former, infinite set has a smaller Kolmogorov complexity. The reason is that in any reasonable programming language, programs that give the same output regardless of the input are shorter than programs that give different outputs depending on which input they receive; one robust fact about computational complexity is that constant functions are easy to compute.[21]

We might dismiss this point as merely an artifact of the particular way I've picked of formalizing the issue, but this would be too quick, as there are intuitive analogs of the formal point. Of the following two rules, the first is easier to remember:

1. Say "yes" in response to any question of the form "Do you know^n that P?".
2. Say "yes" in response to any question of the form "Do you know^n that P?" for $n \leq 13$, but say "no" in response to any question of that form for $n \geq 14$.

Remembering the second rule requires remembering where the crucial cutoff point is, while remembering the first rule does not. The more general

[20] Not every set has a finite Kolmogorov complexity—undecidable sets don't, since for an undecidable set, there is no program that tests for membership in the set. The Kolmogorov complexity of such sets is defined to be infinite. See Li and Vitányi (2008, p. 102).

[21] In treating the infinite set as one whose membership testing program computes a constant function, I'm assuming that the programs we're discussing only take inputs of the form "$K^n p$." This assumption involves little loss of generality, since allowing the programs to take other inputs would just involve adding a fixed, finite length—that involved in testing whether or not an input is of the form "$K^n p$." Once we move to this setting, there's no guarantee that the infinite set is less complex than *all* finite initial subsets, but we still have a guarantee that as n increases, eventually the complexity of the set consisting of the first n members of the sequence will exceed the complexity of the infinite set.

point is that if we think of the complexity of a set of sentences in terms of the complexity of the rule required to determine which sentences are in the set, then there's no general presumption that infinite sets will be more complex than finite sets. A rule that lets in every member of some infinite set may be simpler—because it doesn't need to make distinctions—than a rule that lets in only some finite subset.

Note that infinite sets can be simple even when their members increase in complexity without bound: the set of all natural numbers has a low Kolmogorov complexity, even though the Kolmogorov complexity of individual natural numbers increases without bound (Li and Vitányi, 2008, p. 116).[22] The shortest program that outputs 13378937859424530467780 76549298239455875645 78 on a null input is longer than the shortest program that tests for membership in the set of even numbers;[23] there's no contradiction in a simple set having complex members. This corresponds to the intuitive point that it's much easier to remember to say "yes" when presented with any even number—you can just check the last digit—than it is to remember to say "yes" when presented with the number above, and "no" when presented with any other number.

Once the present approach to thinking about complexity and epistemic iteration is adopted, we may start to wonder whether *opponents* of epistemic iteration principles place unrealistic computational constraints on epistemic subjects. While Greco and Bergmann seem to think our knowledge and justification give out quite quickly, if there are independent reasons for denying this, then the opponent of epistemic iteration principles may have to attribute a high but finite number of iterations of knowledge or justification to epistemic subjects.[24] And as we've already seen, this plausibly involves *more* complex information storage than is posited by the defender of epistemic iteration principles.

So far I've focused on discussing cases that it's intuitively natural to describe as counterexamples to introspection principles, and have argued that they are best interpreted in other terms. In the next section I'll focus on a cluster of more theoretical challenges to introspection principles. In each case, we are faced with a putative forced choice between abandoning introspection principles on the one hand, or embracing an untenable skepticism on the other. I'll argue

[22] I don't mean by this that it *monotonically* increases. Just that for any n, there is a natural number of Kolmogorov complexity $> n$.

[23] Or at least, if that's not true for the particular number above, it certainly is true for some larger patternless number.

[24] And as I'll discuss in the next chapter, I *do* think there are independent reasons to deny that our knowledge and justification give out after just a few iterations; I think accepting such a view would make it hard to make sense of our behavior in cases of (putative) common knowledge.

that in each case, the "skepticism" at issue will look unobjectionable if we are contextualists of almost any stripe—certainly if we are model contextualists.

6.3 Skepticism

Dorr et al. (2014) describe an example they take to make trouble for the following, extremely plausible principle:

FAIR COINS If you know that a coin is fair, and for all you know it is going to be flipped, then for all you know it will land tails.

The example is as follows:

1000 fair coins are laid out one after another: $C_1, C_2, \ldots, C_{1000}$. A coin flipper will flip the coins in sequence until either one lands heads or they have all been flipped. Then he will flip no more. You know that this is the setup, and you know everything you are in a position to know about which coins will be flipped and how they will land. In fact, C_2 will land heads, so almost all of the coins will never be flipped. In this situation it is plausible that, before any of the coins are flipped, you know that C_{1000} will not be flipped—after all, given the setup, C_{1000} will be flipped only in the bizarre event that the previous 999 fair coins all land tails. It follows that there is a smallest number n such that you know that C_n will not be flipped. But then C_{n-1} is a counterexample to FAIR COINS. (p. 277)

While Goodman and Salow (2018) don't object to this reasoning—they agree it amounts to a compelling argument against FAIR COINS, they do suggest the possibility of some damage control. While we can't salvage FAIR COINS, we *can* salvage WEAK FAIR COINS:

WEAK FAIR COINS: If a coin is fair and will be flipped, then for all you know it will land heads.

There can be a first coin you know won't be flipped consistent with WEAK FAIR COINS, so long as the previous coin isn't flipped.[25] E.g., suppose C_i is the

[25] In this respect, WEAK FAIR COINS is reminiscent of the "Margin For Error" principles defended in Williamson (2000). Though Goodman and Salow (2018) do explicitly discuss the differences between the principle they suggest and the ones in Williamson (2000).

last coin flipped. Weak Fair Coins entails that, for all you know, C_i will land tails, and hence, for all you know, C_{i+1} will be flipped. But while Fair Coins then entails that, for all you know, C_{i+1} will land tails, Weak Fair Coins does not. It's consistent with Weak Fair Coins that C_i is the last coin flipped, and for all you know, C_{i+1} will be flipped, but that you know it won't land tails, and so you know that C_{i+2} won't be flipped.

But drawing this distinction requires rejecting KK. While Weak Fair Coins allows that there can be a first coin you know won't be flipped, you can't know *of* this coin that it is the first you know won't be flipped. Goodman and Salow (2018) explain how the distinction between Fair Coins and Weak Fair Coins collapses, given KK, as follows:

> For suppose you know that C_i won't land tails. By KK, you can know that you know this. Plausibly, you can also know Weak Fair Coins. But then you can deduce that C_i won't be flipped from the known facts (i) that you know that C_i won't land tails and (ii) that Weak Fair Coins is true. This would seem to allow you to know that C_i won't be flipped. So C_i is no counterexample to Fair Coins. Since C_i was chosen arbitrarily, this suffices to establish Fair Coins. (p. 184)

While Goodman and Salow (2018) do go on to suggest a potential out for the KK sympathizer, I'm inclined to get off the boat much earlier and reject the assumption that there *is* a first coin you know won't be flipped. Rather, it strikes me as more natural to model you as knowing the setup, and then knowing the probabilistic claims that follow from that, but not more. The probability that C_{1000} won't be flipped is $1 - \frac{1}{2}^{999}$, which is close to but less than 1, and so not known. Dorr et al. (2014) worry that this approach is untenably skeptical. If we can't know that C_{1000} won't be flipped, then almost all knowledge of the future is undermined, since so much of the future depends on chancy processes relevantly similar to the series of coin flips in their example.[26] But I hope, in light of the discussion in Chapter 4, it's clear why this concern doesn't much move me. Treating a claim as merely highly probable but not certain in one context doesn't preclude ignoring possibilities where similar claims are false and thereby treating them as certain in another. Just about any claim that we'll ordinarily model ourselves and others as knowing is one that, in the right context, it will instead be attractive to model us as being merely highly

[26] They discuss an example involving a tree shedding its leaves, but cases can be easily multiplied. See, e.g., Vogel (1999), or Hawthorne (2004), both of whom discuss a variety of cases where it's natural to describe us as knowing facts that it's also natural to think of as merely highly probable but not certain.

confident in. So modeling ourselves as failing to know that C_{1000} won't be flipped doesn't commit us to never modeling ourselves as knowing *anything* about the future. At worst, it commits us to being prepared to retreat from claims we ordinarily take ourselves to know once we start taking seriously certain outlandish but possible scenarios. But if we are already contextualists, this commitment shouldn't seem unwelcome or awkward.

A similar dialectic arises concerning the surprise examination paradox. Williamson (2000) argues that an attractive treatment of it involves appealing to margin-for-error principles that are incompatible with KK. In the course of motivating his treatment of the paradox, he appeals to a similar anti-skeptical principle to the one used by Dorr et al. (2014) to rule out a family of alternative treatments that involve no KK-violation. But, as above, if we are contextualists about knowledge, the species of "skepticism" entailed by KK-preserving treatments of the surprise exam needn't concern us. I'll explain below.

The surprise examination paradox involves a group of students who are told by their teacher that there will be a surprise exam this month, and in which the teacher clarifies that by "surprise" he means that on the morning of the exam, the students won't know that the exam will be given on that day. The students reason as follows. It can't be on the last day of the month, since if they get to the last day of the month without already having had the exam, the students will know there's only that day left, in which case they'd know it was coming on that day, and hence it wouldn't be a surprise. So they rule out the last day. But then they reason that once the last day of the month is ruled out as a possibility, the same argument they just gave will *also* rule out the possibility of the exam being given on the second-to-last day. They continue to rule out every day of the month as a possibility for the exam. In fact, the exam comes some time in the middle of the month, and they're surprised. Where did their reasoning go wrong?[27]

While explaining Williamson's treatment in depth would take us somewhat far afield, the basic idea involves allowing that there can be a last day on which, for all the students know, the exam will be offered, so long as the students don't know *of that day* that it is the last day when, for all they know, the exam will be offered. In Williamson's model, for the students to know the exam won't be on the last day, it must be true that the exam won't be offered on the

[27] The original text of this footnote stated that the surprise exam paradox was introduced in the March 1963 edition of Martin Gardiner's "Mathematical Games" column in *Scientific American*. I then learned that Quine (1953) discussed it well before that, claiming that "it has had some currency from 1943 onward." Wherever it's from, it has generated a great deal of literature spanning multiple disciplines.

penultimate day. But it doesn't follow that the students can *know* that it won't be on the penultimate day. If the students not only know, but know that they know that it won't be on the last day, then it follows that they can know it won't be on the penultimate day, and it also follows that in fact it won't be offered on the prepenultimate day. But again, this last claim needn't be known by the students. In short, while the student's reasoning *would* go through given the KK principle, without it, it is liable to fail at some step, at least if the month is long enough. Which step it fails on will depend on how many iterations of knowledge they initially had for the claim that the exam won't be on the last day.

I haven't provided enough context to make it clear why this might seem like an attractive treatment of the case. In Williamson's own presentation, it is presented as a natural generalization of some independently attractive models of the limits of our perceptual knowledge. But, as in the case of FAIR COINS it seems to me there is a flatfooted alternative approach—denying that the students can know that there will be a surprise examination—whose putative drawbacks look much less concerning once we adopt the modest modeling perspective urged in this book.

A representative example of that approach is provided by Hall (1999).[28] In his model, after hearing the teacher's announcement the students are highly confident that there will be a surprise exam in the following sense: they are highly confident there will be an exam, and highly confident that on the morning of the exam they won't be highly confident that the exam will be offered that day. But they don't know these things (at least if knowledge requires probability 1). The explanation for why their reasoning goes wrong is that, conditional on the exam's not having been offered before the last day, the students are not highly confident of both these things. And this is explained in a natural way as the upshot of conditionalization—with each day that passes, they become a little bit less confident that there will be a surprise exam. If they get all the way to the last day, then depending on their priors, they may end up confident that there will be no exam at all, or perhaps instead that there will be an exam which won't come as a surprise.

Williamson rejects treatments like these on anti-skeptical grounds:

> Advance knowledge that there will be a test, fire drill, or the like of which one will not know the time in advance is an everyday fact of social life, but one denied by a surprising proportion of early work on the Surprise Examination. Who has not waited for the telephone to ring, knowing that it will do so within a week and that one will not know a second before

[28] The discussion in Quine (1953), while much briefer, is in a similar spirit.

> it rings that it will ring a second later? Any adequate diagnosis of the Surprise Examination should allow the pupils to know that there will be a surprise examination. (2000, p. 139)

As common-sensical as this sounds, to my own ears it seems a lot less so when we remember that knowledge is being (implicitly) sharply contrasted with high but non-maximal probability. We already saw in Chapter 4 that allowing for the possibility of a claim's being defeated by future evidence requires modeling the claim as initially uncertain. The surprise examination is merely a special case of that more general phenomenon. In describing the situation of the students, we want to be able to accommodate the fact that going almost the entire month without an exam would rationally undermine their confidence in the teacher's announcement; that's what Hall's model is designed to do. But then to do that, we need to model them as initially uncertain—albeit highly confident—in the content of that announcement.

And as we've already seen, this doesn't commit us to a general skepticism. Modeling the students as not knowing that there will be a surprise exam doesn't commit us to never modeling anyone as knowing anything. Instead, it commits us to modeling people as not knowing claims when we're trying to make specific allowances for future evidence that would undermine those claims. Because just about any claim is subject to such undercutting, just about any claim is one that, under certain circumstances, we'll sensibly model people as failing to know.

I hope the above considerations may also shed some light on a longstanding disciplinary divide. As I mentioned earlier in this chapter, economists and computer scientists interested in knowledge typically use modeling frameworks in which agents are both positively introspective ($K \rightarrow KK$) and negatively introspective ($\sim K \rightarrow K\sim K$).[29] Philosophers tend to regard these modeling frameworks as highly idealized, admittedly useful for applications involving the prediction of choice behavior or the design of distributed systems, but still getting fundamental facts about the structure of knowledge *itself* wrong.[30] However, from within the modest modeling perspective urged in this book—one where, as we saw in Chapter 1, modeling frameworks aren't the kinds of things that are assessable for truth or falsity—it's not so straightforward to draw the distinction between a framework for modeling knowledge that is useful for a variety of applications, and one that accurately

[29] See, e.g., Samuelson (2004) and Fagin et al. (2003).

[30] See, e.g., Williamson (2000, p. 122), Lederman (2018a, pp. 17–18).

reflects the structure of knowledge itself. In the remainder of this chapter I'll explore that question, with an eye towards providing a more explicit and principled defense of what I take to be the implicit practice—modeling knowledge as positively and negatively introspective without feeling too much guilt—of many economists and computer scientists.

What sort of relationship between introspective and non-introspective frameworks should philosophical critics of introspection principles be understood as positing? Clearly it's not similar to the relationship between, say, Ptolemaic and Copernican frameworks for modeling planetary motion. If that were true—if non-introspective frameworks could explain everything introspective frameworks could explain while being more elegant and tractable at the same time—then the use of introspective frameworks outside of philosophy wouldn't have persisted this long. A more promising analogy, used by Williamson, is between models that incorporate introspection principles on the one hand, and models that ignore friction or air resistance on the other.[31] On closer inspection, however, I don't think the analogy holds up. Let's see why not.

Let's focus on the case of air resistance. Just what is the relationship between models of falling objects that incorporate resistance, and those that don't? At the risk of oversimplifying, I'll attempt the following survey. Models that don't incorporate resistance are easier to use. But even in the best cases for models that leave out air resistance, models that incorporate it are accurate within narrower margins of error (e.g., in their predictions about rates of descent). Moreover, models that incorporate air resistance have broader scope. Models that incorporate air resistance can be straightforwardly generalized to situations where the density of the object is comparable to the density of the medium through which it is moving (e.g., tapioca balls in bubble tea), while models that ignore resistance can't handle such cases. Cases where two modeling frameworks differ in this way—one framework is less tractable than the other, while also being strictly more accurate in its predictions and having broader explanatory scope—strike me as the paradigm cases for our being able to identify local features of modeling frameworks as "mere idealizations." Does the relationship between non-introspective and introspective frameworks for modeling knowledge fit into this paradigm?

In some ways it does. Non-introspective frameworks are much more cumbersome to use to represent interpersonal knowledge; it's no accident that the models in Williamson (2000) and the literature it has inspired are almost

[31] See (Williamson, 2000, p. 122).

without exception models of a single agent's knowledge. Perhaps this reason alone explains why such models have not found much uptake in disciplines where knowledge is typically studied in the context of strategic interaction between multiple agents. But in my view, the analogy stops there. While non-introspective frameworks have comparable drawbacks to frameworks incorporating resistance—they are less tractable, at least when modeling multiple agents—it's not at all clear that they also have analogous advantages.

Do non-introspective frameworks make predictions that are accurate within narrower margins of error than introspective frameworks? This question is perhaps more natural for economists—who are likely to see models of knowledge as in the business of making predictions about strategic behavior—than for philosophers, who tend to see looser connections between knowledge and choice. At the very least, it's a contentious question what, if anything, to count as the "predictions" of models of knowledge—comparable to the rates of descent produced by models of free fall—such that we could then ask whether non-introspective models make strictly more accurate such predictions. My suspicion is that philosophers who defend such models will tend to give a trivial answer here; "non-introspective models make the more accurate 'prediction' that you can know without knowing that you know, or fail to know without knowing that you fail to know." From the perspective I've been urging in this book—especially in Chapter 2—that should sound unsatisfying. Only when we're trying to limn the fundamental structure of reality should we be content with a defense of a model that says, effectively: "that's just how things are!" Whenever we're modeling reality in non-fundamental terms, we can rightly ask for some non-trivial payout of doing so. So it seems to me reasonable to ask, not just that non-introspective models of knowledge be more accurate *about knowledge* than their introspective cousins, but that they be more accurate about something else, such as patterns in choice behavior. I suspect this will sound objectionably "behavioristic" or "operationalistic" to most philosophers. In my view, however, it's the natural consequence of an independently attractive view about reduction and emergence quite generally, rather than a distinctively objectionable view about knowledge as such.[32]

[32] Kevin Dorst's "Rational Polarization" is an important example of a philosopher proposing a non-introspective framework for modeling knowledge, but where the methodological criticism I'm discussing here doesn't apply. In that paper he offers a positively but not negatively introspective framework for modeling knowledge. As far as I can tell, he would be happy with the idea that his models should be judged on their ability to make plausible predictions and/or recommendations about phenomena *other* than knowledge itself, such as choice behavior; that is, in fact, the kind of phenomenon he appeals to in arguing for the fruitfulness of his models. While I do have my disagreements with Dorst—I tend to think the particular kinds of reflection violations his models generate are neither a good descriptive match for our actual behavior, nor do they make for attractive

If we *do* focus on predictions about choice behavior, the situation isn't at all analogous to that with models that incorporate air resistance, which make accurate predictions but can be cumbersome to use. Rather, as we saw in §6.1 and §6.2, non-introspective models, even when applied to examples where they are quite tractable, seem to make a wide variety of *less* accurate predictions than their introspective cousins.[33] Do non-introspective models have broader scope than introspective models? That is, are there situations where they shine while their introspective cousins fall flat (analogous to the case of tapioca balls in bubble tea)?

This is a difficult question, whose full resolution is beyond the scope of this book. The cases I discussed at the beginning of this section—FAIR COINS and the surprise exam—were meant to represent some contenders for illustrations of the broader scope of non-introspective modeling frameworks. But I hope that discussion made plausible that, at best, those cases are a wash; introspective frameworks that make "skeptical" concessions can handle the cases just as well as non-introspective frameworks. Moreover, I tend to think that understates the advantage that "skeptical" introspective frameworks have; I suspect if we asked questions about how, e.g., the students in the surprise exam case would or should behave, we'd see that the "skeptical" introspective models plausibly do better than their non-introspective, "anti-skeptical" cousins. But those cases aren't the only ones in which non-introspective frameworks have been thought to be fruitful. In particular, a full accounting of the comparative costs and benefits of introspective and non-introspective frameworks for modeling knowledge would have to take account of the phenomenon of vagueness. Williamson (1994) offers an "epistemicist" treatment of vagueness, in which vagueness is a species of ignorance. In his view, the vagueness of "tall" consists in our inability to know just how much height one needs to be tall. Some of the main virtues of his account crucially rely on failures of introspection principles. In models incorporating positive and negative introspection principles, we'd know, for each height in nanometers, whether we know that having that much height is enough to be tall. This, Williamson argues, amounts to failing to capture higher-order vagueness; not only is it vague how much height is necessary to be tall, it's also vague where

ideals—they're much more in the weeds, and less at the level of methodology, than my disagreements with the philosophers I've quoted more extensively in this section.

[33] Though I should note that I take this claim to be tentative. Schulz (2017) offers a generalization of decision theory meant to allow higher-order intrapersonal knowledge to do non-trivial work. While I'm not yet convinced of its merits—among other concerns, I suspect it can't be smoothly generalized to multi-agent cases, for reasons similar to those I discuss in §1.3 of Greco (2014a)—if such models do prove to be fruitful for a wide range of applications, I'd likely reverse my judgment in this section.

the vagueness begins. If vagueness is ignorance, then we can only capture higher-order vagueness if we are ignorant about the extent of our knowledge (ignorance).[34] Modeling such ignorance requires rejecting both positive and negative introspection principles.

Just how powerful a reason to favor non-introspective frameworks is this? The answer depends on just how compelling are the reasons to favor an epistemicist account of vagueness, which as noted is beyond the scope of this book to address.

So where does this leave the claim that modeling frameworks that incorporate introspection principles are more idealized and less accurate than modeling frameworks that eschew them? I hope that claim looks considerably thornier than it might initially have seemed. The picture I've tried to paint in this chapter is one on which formal and informal modeling frameworks that "conflate" intrapersonal level distinctions—e.g., frameworks that conflate an agent's beliefs with her beliefs about her beliefs, or with her beliefs about what she should believe, or which conflate her knowledge with her knowledge of her knowledge—are much more defensible than might at first appear. These conflations do real explanatory work—in capturing facts about rational choice (e.g., that aviophobe Matt can be expected to seek therapy), conversational dynamics (e.g., that Bob must be trolling), and the marked contrast between the smoothness of higher-order interpersonal attitude ascriptions and the roughness of otherwise similar higher-order intrapersonal attitude ascriptions. And the most obvious apparent drawbacks of the frameworks—their apparent inability to describe cases of first-order but not second-order belief and knowledge—can be addressed by interpreting those cases via a fragmentationist/modest modeling lens. But they may have serious drawbacks when it comes to modeling vagueness. Our situation is somewhat analogous to the ones Kuhn (1962) describes as the prelude to scientific revolutions. Introspective frameworks more neatly capture some phenomena, while non-introspective frameworks are best for others, and their respective partisans tend to disagree about which phenomena are more central and which can be set aside as curiosities, or handled with ad hoc adjustments. But unlike Kuhn, who saw such situations as unstable,[35] to a modest modeler there's no obvious reason why this can't be an equilibrium. While it would be *nice* for one approach to ultimately subsume the other, or for there to emerge some

[34] The incapacity of alternative theories of vagueness to smoothly accommodate higher-order vagueness is one of the recurring themes in Williamson (1994).

[35] See, e.g., Kuhn (1962, pp. 162–3).

Hegelian synthesis with the virtues of both and the vices of neither, modesty about epistemological modeling would suggest that this isn't inevitable.

As already alluded to, outside of philosophy much of the interest in higher-order knowledge concerns interpersonal knowledge, and especially common knowledge. In the next chapter I'll discuss a recent argument to the effect that we never have common knowledge. As in the case of introspection principles, I'll argue that modeling agents as (potentially) having common knowledge looks much more defensible when we are modest about modeling.

7
Modeling Common Knowledge

In this chapter I'll apply some of the themes in the book thus far to the topic of higher-order *inter*personal knowledge, and especially common knowledge. In §7.1 I'll introduce the idea of common knowledge and draw some connections between common knowledge and the level "conflations" I defended in the previous chapter. In §7.2 I'll turn to a recent argument due to Harvey Lederman (2018b) to the effect that common knowledge is unattainable. I'll suggest that Lederman's argument looks much less compelling once we are modest about models of knowledge.

7.1 Introducing Common Knowledge[1]

To a first approximation, a proposition P is "common knowledge" among a group if everyone in the group knows that P, everyone knows that everyone knows that P, everyone knows that everyone knows that everyone knows that P, and so on *ad infinitum*. The concept of common knowledge has been used in a wide range of disciplines to do diverse explanatory jobs. Philosophers have appealed to common knowledge to explain linguistic meaning, conventions, and various forms of coordination more generally,[2] linguists have appealed to common knowledge to characterize the sort of shared information necessary for certain sorts of speech acts to be felicitous,[3] much of game theory involves reasoning about how rational agents will behave when certain facts (e.g., facts about the structure and payoffs of some game that they are playing) are common knowledge,[4] and computer scientists often use the concept of common knowledge in theorizing about distributed systems.[5]

[1] Section §7.1 is based on §2.1 of my (2015).

[2] See, e.g., Schiffer (1972), Lewis (2002/1969), and Heal (1978), respectively.

[3] Clark and Marshall (1981).

[4] Aumann (1976) is the locus classicus for the use of the concept of common knowledge in the context of game theory. Though see Bicchieri (1989) for criticism of the idea that predictions in game theory depend on agents having common knowledge of the structure of the game. See also Crawford et al. (2013) for a recent survey of work in game theory that models strategic behavior without appealing to common knowledge, and Lederman (2018a, b) for philosophical discussions of how common knowledge assumptions can (and should, according to the author) be relaxed in game theory.

[5] See Fagin et al. (2003).

Idealization in Epistemology: A Modest Modeling Approach. Daniel Greco, Oxford University Press. © Daniel Greco 2023.
DOI: 10.1093/oso/9780198860556.003.0008

In this section, while I'll try to briefly give a sense of why so many theorists have found common knowledge to be such a useful concept, my main focus will be on the relationship between allowing for common knowledge on the one hand, and "conflating" intrapersonal level distinctions on the other. In particular, I'll argue that some of the most natural explanations of how it is possible for finite creatures to attain common knowledge—explanations that, in one form or another, have been offered by many of those who theorize about common knowledge—rely on frameworks that fail to make intrapersonal level distinctions. This provides a kind of defense of such frameworks. If a wide range of phenomena are best explained by positing common knowledge, but common knowledge is best represented in modeling frameworks that ignore intrapersonal level distinctions, then we have some reason for using such frameworks.

7.1.1 Common Knowledge and Coordination

Consider the contrast between the following two cases:[6]

> Public Announcement: A professor tells her class that they will play the following game. Without communicating with one another in any way, each student in the class will write down the name of a US state on a piece of paper. If all students write the same state name, with the exception of the name of the state the class is taking place in, the students will each receive $10. If any two students write down different state names, or if they all write down the name of the state the class is taking place in, no prize money will be awarded. Before handing out the pieces of paper, the professor tells the class that she grew up in Maine (which is not the state the class is taking place in), and that it is lovely in the fall.
>
> Private Information: Just like the previous case, except instead of publicly announcing that she grew up in Maine, the professor whispers the following to each student privately as she hands out the pieces of paper: "while I'm not telling anybody else this, I'd like you to know that I grew up in Maine, and it is lovely in the fall."

[6] The strategy of introducing the concept of common knowledge by contrasting cases in which some fact is common knowledge among a group, and cases in which it is not, even though it is known by each member of a group, or even known to be known (or known to be known to be known), is common. See Heal (1978), Clark and Marshall (1981), or any discussion of the "muddy children" puzzle (e.g., that in Fagin et al. 2003) for some paradigm examples.

Plausibly, the students will behave quite differently in the two cases. In the former case, it's likely that the students will each write down "Maine," since Maine has been made publicly salient. They'll probably win the prize. In the latter case, however, while this is possible, it is much less likely. Each student will think that while Maine is salient to *her*, it isn't salient to the other students, so there's no special reason to think that they'll write down "Maine." And if there's no special reason to think that the *other* students will write down "Maine," none of the students will regard *themselves* as having any special reason to write down "Maine." The students will probably try to coordinate on some other salient state—New York? California? They are unlikely to win the prize.

Supposing that the students manage to coordinate on Maine in the former case, what it is it about the case that explains and justifies their each writing down the same state name? It's *not* that everybody knows that Maine has been singled out by the professor in some way—that is true in PRIVATE INFORMATION as well. A tempting thought is that in PUBLIC ANNOUNCEMENT but not PRIVATE INFORMATION, not only does everybody know that Maine has been singled out, but it's *also* the case that everybody knows that everybody knows that Maine has been singled out, and that this is the key distinction that explains why everybody is likely to write down Maine (and has a strong reason to do so) in PUBLIC ANNOUNCEMENT, but not in PRIVATE INFORMATION. But this distinction won't work either, since we can construct a variant of PRIVATE INFORMATION in which everybody knows that Maine has been singled out, *and* everybody knows that everybody knows that Maine has been singled out, but in which students still lack strong reasons to coordinate, and are unlikely to do so:

> MORE PRIVATE INFORMATION: Just like the previous case, except this is what the professor whispers: "I'm privately telling everybody in the class that I grew up in Maine and that it's lovely in the fall. However, you're the only one who I'm telling *that* I'm telling everyone. Each other student thinks that she's the only one who knows that I grew up in Maine."

We already established that in PRIVATE INFORMATION, students are not particularly likely to successfully coordinate. But in MORE PRIVATE INFORMATION, each student thinks that the *other* students take themselves to be in a situation like PRIVATE INFORMATION, and so to be unlikely to pick Maine as opposed to any of various other potentially salient states. So if students are unlikely to coordinate on Maine in PRIVATE INFORMATION, they're also unlikely to coordinate on Maine in MORE PRIVATE INFORMATION. But in MORE PRIVATE INFORMATION, each student not only knows that Maine has been singled out, but also knows that the other students know that Maine has been singled out.[7] It's easy to see how to keep going constructing variants of PRIVATE INFORMATION; with another iteration we could get a case where the students not only all know that they all know that Maine has been singled out, but they *also* all know that they all know that they all know that Maine has been singled out, and still coordination will be unlikely.[8]

Many writers have thought that the key difference between PUBLIC INFORMATION and any of the cases in the hierarchy starting with PRIVATE INFORMATION and MORE PRIVATE INFORMATION is that in PUBLIC INFORMATION, the students' knowledge doesn't give out at *any* level of iteration. That is, they all know that Maine has been singled out, they all know that they all know this, they all know that they all know that they all know this... and so on for as many iterations of "they all know" as you like. If any level of iteration doesn't hold (as the second fails in PRIVATE INFORMATION, and the third fails in MORE PRIVATE INFORMATION), the students will lack strong reasons to write down Maine, and they are unlikely to successfully coordinate. A weaker version of this idea is that, even if the students don't literally have each knowledge state in the infinite iterated hierarchy, they have some sort of evidential basis from which it's clear that—with sufficient time and logical acuity—each level of the hierarchy could be generated. That is, there is no epistemologically principled barrier to their achieving arbitrarily many iterations of knowledge, even if, in practice, they are unlikely to ascend all that far up the hierarchy.[9] This sort of view will still contrast cases of common knowledge, in which there are no principled barriers to ascending all the way up the hierarchy, with cases like PRIVATE INFORMATION, in which there are such barriers.

Before moving on, I'd like to echo a qualification that is standard in work on common knowledge.[10] To the extent that common knowledge is thought of as

[7] Or at least the students rationally believe these things—maybe the fact that what the professor says is partially a lie is enough to prevent the students from coming to know the true parts.

[8] We can argue for this by induction. In the first case, they won't coordinate. In each successive case, each student thinks that everybody else takes herself to be in the previous case. So if the students are unlikely to coordinate in case n, then they are also unlikely to coordinate in case $n + 1$, when each student thinks that the other students think that they are in case n. The reasoning is very similar to the reasoning in Rubinstein (1989), or the explanation in Fagin et al. (2003) of why common knowledge can't be generated by communication in systems with unreliable messaging.

[9] E.g., according to David Lewis, common knowledge involves an infinite hierarchy of reasons to believe (i.e., if it is common knowledge between us that P, we each have reason to believe that P, and reason to believe that we have reason to believe that P, and reason to believe that...), but we typically only generate finitely many higher-order expectations on the basis of these reasons.

[10] See, e.g., Heal (1978, p. 116).

a foundation for certain sorts of coordination, the label "common knowledge" is somewhat misleading, as it is possible to coordinate around information that is not known. Suppose we are making plans to see a movie, and I make the following proposal to a group of friends: "Let's meet at the theater fifteen minutes before the movie starts." Suppose further that, while we commonly take for granted that the movie starts at eight, it actually starts at nine. In such a case, one should predict that we'll all show up to the theater fifteen minutes before eight—in one sense, we'll succeed in coordinating.[11] To the extent that one is sympathetic to common knowledge explanations, it's natural to think that explaining how we managed to coordinate will require appealing to some sort of common-knowledge-like attitude that we bear towards the claim that the movie starts at eight. But that attitude cannot literally be common knowledge—we don't commonly know that the movie starts at eight, as the movie does not start at eight. I take the lesson of the example to be that questions about common "knowledge" really arise for *any* sort of attitude—belief, acceptance, presupposition, etc.—that can figure in explanations of coordinated action. Moreover, if explaining how we attain common knowledge requires positing an iteration principle for knowledge—a view that I'm about to discuss—then we'll also have to posit iteration principles for other attitudes that can play similar roles in explanations of coordinated action.

7.1.2 Common Knowledge and Epistemic Levels

Assuming that certain sorts of coordination really are best explained by appeal to common knowledge—an assumption we'll return to later—how do finite agents manage to attain common knowledge? Clearly not by separately acquiring and/or storing each level of an infinite hierarchy of knowledge states—that would require prohibitively much time and memory. Rather, there must be some finite basis from which each level of the hierarchy is (or can in principle be) generated.[12] What might that basis be?

Many writers have proposed some version of the idea that groups attain common knowledge when not only do they each know that P, but it is clear to each of them that their epistemic situation vis-á-vis P (and what any of

[11] Of course, in another, we'll fail, as we were trying to show up fifteen minutes before the movie started.

[12] While there is much that those who work on common knowledge disagree about, the requirement that common knowledge have a finite basis if it is to do genuine explanatory work is, as far as I can tell, a point of consensus.

them are in a position to know concerning P) is symmetrical.[13] For example, suppose that while you and I are having a conversation, a goat walks into the room and loudly bleats. Not only will it be clear to each of us that there's a goat in the room, it will also be clear that we're equally well positioned to know that there's a goat in the room. I'll be happy to ask "how did that thing get here?" without worrying that you don't know what I'm referring to. How does the fact that our situation is epistemically symmetrical allow us to achieve common knowledge? If we allow ourselves the KK principle, it's quite straightforward: the combination of a symmetry assumption to the effect that we're in a position to know the same things concerning the existence of the goat and our knowledge of it, and the KK principle, will generate each level in the common knowledge hierarchy, as follows:

1. I know there's a goat in the room.
2. You know that there's a goat in the room (by symmetry).
3. I know that I know that there's a goat in the room (by 1, and KK).
4. You know that I know that there's a goat in the room (by 3, and symmetry).
5. You know that you know there's a goat in the room (by 2, and KK).
6. I know that you know there's a goat in the room (by 5, and symmetry).

   ⋮

In the absence of something like KK, however, we can't get past step 2.[14] From the fact that I know that P, and that our situations are symmetrical, it won't follow that you know that I know that P—it will only follow that you know that P. Suppose that each extra iteration of higher-order intrapersonal knowledge is non-trivial. Then I might know that P, without knowing that I know it. If our situations are symmetrical, you will also know that P without knowing that you (or I) know it. In such a case, even though our situations are symmetrical, we won't be in a position to generate the common knowledge hierarchy symmetry will guarantee that however many iterations of knowledge I have, you have just as many, but it won't guarantee that we are in a position to go all the way up. In particular, if I only have one iteration of knowledge—I know, but do not know that I know—then the fact that our situations

[13] See, e.g., Heal (1978), Clark and Marshall (1981), and Gilbert (1989).

[14] For this reason, I take it that the similar derivation in Heal (1978, p. 126) should be understood as implicitly relying on something like KK.

are (knowably) symmetrical will guarantee that you also have a mere single iteration of knowledge, and that neither of us knows that the other knows.

The connection between the possibility of common knowledge on the one hand, and modeling frameworks on which higher-order intrapersonal knowledge is trivial, is even stronger than the above might suggest; it's not just that such frameworks open up one route—the symmetry based route—to achieving common knowledge. If this were the extent of the connection between such frameworks and common knowledge, there still might be some other route to common knowledge that didn't require "conflating" epistemic levels. However, views that make such level-distinctions typically have, as a side effect, the consequence that there are always principled obstacles to achieving arbitrarily many iterations of *intra*personal knowledge—a single agent is never in a position to know that she knows that she knows…for arbitrarily many iterations. A fortiori, groups of agents are never in a position to achieve common knowledge—if I can't know that I know that I know that P, then we can't know that we know that we know that P.[15]

Why do views on which KK fails typically entail that there are always principled obstacles to achieving arbitrarily many iterations of *intra*personal knowledge? The basic idea is that such views entail that each extra iteration of knowledge is more epistemically demanding than the last, and that achieving arbitrarily many iterations of knowledge of some fact would require an impossibly strong epistemic position vis-á-vis that fact.[16] For example, on a Williamsonian (2000) approach to higher-order knowledge, seeing a goat from a distance in poor lighting might put one in a strong enough epistemic position to know that there is a goat present, without putting one in a strong enough position to know that one knows.[17] Seeing a goat up close in good lighting would put one in a stronger position—perhaps enough to achieve four or five iterations of knowledge—but still not an arbitrarily strong position, and so still not enough to achieve arbitrarily many iterations of knowledge that there is a

[15] See Hawthorne and Magidor (2009). Though Goldstein (2022) is an important counterexample. He proposes a framework in which while knowledge doesn't automatically iterate—you can have K without KK—once you achieve some number of iterations of knowledge, maybe just two, *then* further iteration comes for free. So he wants to thread the needle between accepting KK and the possibility of "omega knowledge" (i.e., infinitely iterated knowledge) on the one hand, and denying KK as well as the possibility of Omega Knowledge on the other. He thinks KK fails, but omega knowledge is possible nevertheless. I'm inclined to think that, in Goldstein's models, pretty much all the interesting work—in explaining learning, action, and assertion—is done by omega knowledge, and none of it by regular old, non-iterated knowledge. So I see his models as akin to models that validate KK—because omega knowledge automatically iterates—but with a bit of extra complexity (boring, non-iterated knowledge) that doesn't really pull its own weight.

[16] Greco (2014a) goes into more detail on this point. See also Hirsch and Jenny (2019).

[17] See especially ch. 6 of Williamson (2000), and the discussion of "glimpse" cases.

goat present. Since achieving arbitrarily many iterations of knowledge of a fact would require an impossibly strong epistemic position, none of us are ever in a position to attain common knowledge.

7.2 Lederman's Sailboat

All this is beside the point, however, if common knowledge is unattainable for reasons that have nothing to do with the failure of introspection principles. This is what Harvey Lederman (2018b) argues in a fascinating recent paper. If he's right, then a defense of introspection principles that appeals to their necessity for representing common knowledge is doomed to failure, since common knowledge is unattainable for independent reasons.

Lederman's argument turns on his analysis of the following example:

> SAILBOAT: Roman and Columba are ideal reasoners playing in a game show. Each contestant has a single button on a console in front of him or her. They have an unobstructed view of each other's faces, and of an area in the middle of the stage, where the hosts will place a sailboat. First, the hosts will bring out a toy sailboat (the "test") with a 100 cm tall mast. They will then replace it with a sailboat chosen randomly from an array of sailboats of various sizes. If the mast of the new sailboat is taller than the test and both players press their respective buttons, they receive $1,000 each. If the mast is not taller than the test, and both press, or if only one person presses their button, the person or people who pressed must pay the show $100. Today, the mast of the chosen boat is 300cm tall.

The point of the example is to present a situation in which (1) some fact—in this case, the fact that the mast is taller than 100 cm—is clearly "public" between Roman and Columba in whatever pre-theoretical sense is important in explaining coordination and cooperation, but (2) that fact is *not* common knowledge between them—it is not the case that they both know that they both know that they both know ... [*ad infinitum*] ... that the fact holds.

Lederman appeals to (1) and (2) to argue that the theoretical notion of common knowledge—where the members of a group are modeled as having infinitely iterated, or at least iter*able*, knowledge of what each other know—is inadequate as an analysis of our pre-theoretical notion of public information.

But why accept (2)? What's wrong with modeling SAILBOAT such that it's common knowledge between Roman and Columba that the mast is greater than 100 cm? The problem is that doing so is inconsistent with also validating a principle Lederman calls "interpersonal ignorance":

INTERPERSONAL IGNORANCE For all r, if [the mast] looks to be r cm tall to one of the agents, then for all that agent knows, it looks to be $0.97r$ cm tall to the other.

What's the tension between a model's representing it as common knowledge between Roman and Columba that the mast is greater than 100 cm tall, and also validating INTERPERSONAL IGNORANCE? While Lederman spells out the details, the basic idea is quite simple. Assume that Roman and Columba's probabilistic beliefs about the height of the mast, *and* about the probabilistic beliefs of the other party, are formed by making an estimate, and assuming that both one's own and the other party's estimate will be accurate within some margin of error. Crucially, when your estimate is that the mast is r cm tall, you don't know that it's not r cm tall.

Now, suppose the mast is in fact 300 cm tall, and this is how tall it looks to Roman. So for all he knows, it's 300 cm tall. By INTERPERSONAL IGNORANCE, when Roman's estimate of the mast's height is 300 cm, for all he knows, Columba's estimate is 299 cm. So for all he knows, for all she knows, it's 299 cm. And when Columba's estimate is 299 cm, for all she knows, Roman's estimate is 298 cm. So for all Roman knows, for all Columba knows, for all Roman knows, the mast is 298 cm. With many more steps filling in the ellipsis, you can get the result that for all Roman knows, for all Columba knows, for all Roman knows... the mast is shorter than 100 cm; i.e., it is not common knowledge between them that the mast is taller than 100 cm.

Moreover, Lederman argues that *any* situation in which we might be tempted to claim that some agents have common knowledge of some fact is one in which an argument along the same lines as the one above will be available.[18] The general strategy is to take some putative case of common knowledge, and then describe both the possible states of nature and Roman and Columba's possible perceptual states in very fine-grained terms. E.g., suppose Roman and Columba are standing next to each other, and an adult Golden Retriever is 5 feet in front of them. It's tempting to say that they're in a position to have common knowledge that there's a dog in front of them. But we can find some fine-grained parameters describing their best estimates of the shape, hue, and location of the object in front of them, such that if Roman's

[18] See Lederman (2018b, §3.4).

best estimate of the shape/hue/location is $<s, h, l>$, then for all he knows, Columba's might be $<s \pm k, h \pm k', l \pm k''>$, and likewise for Columba. So by the same argument as before, even if $<s, h, l>$ correspond to paradigm values for a nearby dog and are the true values, and Roman estimates that they are the true values, for all Roman knows, for all Columba knows, for all Roman knows... [insert many steps here]... the true values of the shape, location, and hue of the object in front of them are very far from $<s, h, l>$, and in fact are more like those of a distant rock than a nearby dog. So it's not common knowledge between Roman and Columba that they're looking at a nearby dog, rather than a distant rock.

Lederman's argument, by his own account, "can be seen as an elaboration" of an earlier argument in Fagin et al. (2003). After providing a model of the famous "muddy children" experiment in which the content of a father's announcement becomes common knowledge among a group of children once the announcement is made, they show how if time is modeled at a more granular scale—so that once a child has heard and processed the announcement, for all she knows, the other children may still be parsing it—the content of the announcement never becomes common knowledge. After having offered such a model, they could have said that because time *does* come in very fine-grained increments (or perhaps is continuous), common knowledge is in fact never achievable. But they didn't. Instead, they draw a different conclusion, and one which is much more congenial than Lederman's to the approach advocated in the present book:

> If a model has a coarse notion of time, then simultaneity, and hence common knowledge, are often attainable. In synchronous systems, for example, there is no temporal imprecision. As a result, in our simplified model of the muddy children puzzle, the children do attain common knowledge of the father's statement. As we argued above, however, if we "enhance" the model to take into consideration the minute details of the neural activity in the children's brains, and considered time on, say, a millisecond scale, the children would not be modeled as hearing the father simultaneously. Moreover, the children would not attain common knowledge of the father's statement. We conclude that whether a given fact becomes common knowledge at a certain point, or in fact whether it ever becomes common knowledge, depends in a crucial way on the model being used. While common knowledge may be attainable in a certain model of a given real world situation, it becomes unattainable once we consider a more detailed model of the same situation.

> We are thus led to the question of when we are justified in reasoning and acting as if common knowledge is attainable. This reduces to the question of when we can argue that one model—in our case a coarser or less detailed model—is "as good" as another, finer, model. The answer, of course, is "it depends on the intended application." (2003, §11.4)

Later in this chapter I'll show how, just as in Fagin et al.'s analysis, a less fine-grained model of SAILBOAT would represent it as common knowledge between Roman and Columba that the mast is greater than 100 cm. But before presenting such a model, I want to dwell a bit on the different approaches taken by Fagin et al. on the one hand, and Lederman on the other. Here's how Lederman distinguishes his argument from theirs:

> The most important difference from a formal perspective is that my argument is conducted in the object language, using premises about the agents' knowledge. Unlike these earlier arguments, the object-language argument does not depend on choices about how to model the agents' uncertainty; for example, it does not require directly motivating assumptions about worlds or accessibility relations. Relatedly, the object-language argument shows that the formal result is not an artefact of unintended idealizations implicit in the standard models for knowledge and belief. (2018b)

As I interpret Lederman, he sees Fagin et al. as offering *models* of knowledge, while Lederman instead intends to be directly arguing about knowledge *itself*. He's proposing INTERPERSONAL IGNORANCE not merely as part of an attractive model of SAILBOAT, but simply as a *truth*. From the standpoint of the present book, however, such a distinction looks hard to maintain. Object-language knowledge ascription is part of folk psychology, and we saw in Chapter 3 that folk psychology is fruitfully viewed through the lens of model-building. When we describe people as knowing, believing, or desiring things, we're constructing informal folk psychological models. The sort of modeling that goes on in decision and game theory is a more precise and regimented version of what is fundamentally the same kind of enterprise.

So it's reasonable to interpret an argument to the effect that people never have common knowledge or any approximation thereof as an invitation to engage in model comparison. Are fine-grained models in which common-knowledge is unattainable always superior to coarse-grained models in which it can be achieved? Do they provide better (more accurate? more powerful?

simpler?) explanations of phenomena like, e.g., button-pushing in SAILBOAT, or writing down "Maine" in PUBLIC ANNOUNCEMENT? These are the questions I'll turn to in the remainder of this chapter. But first a qualification. My main aim in this chapter is to show that Lederman's argument isn't obviously compelling. In effect, he's offered the start of a recipe for generating common-knowledge-free models of situations usually modeled with common knowledge. But he hasn't engaged in the process of systematically comparing the explanatory virtues of the common-knowledge-free models to those of the traditional ones. And because of the wide range of applications of common knowledge, I can't really do that here either. My aim is more modest; I hope to show that there's no quick or obvious route to the conclusion that fine-grained, common-knowledge-free models are superior to traditional models that allow for common knowledge.

7.2.1 Granularity

Before presenting a coarse-grained model of SAILBOAT in which Roman and Columba *do* have common knowledge, I want to suggest that the issues here aren't really unique to *common* knowledge. Rather, Lederman's argument has a lot in common with the skepticism of certainty we encountered in Chapter 4, according to which what we ordinarily call knowledge is almost always mere high probability.

It's a familiar point in epistemology that whether we're willing to ascribe knowledge to subjects depends on which possibilities we're considering. We may be happy to say that Susy knows that she's looking at a zebra when the only alternatives we're considering are the other sorts of animals one typically finds in zoos: lions, giraffes, elephants, etc. But once we start considering the possibility that she's looking at a cleverly disguised mule, we may withdraw our attribution of knowledge.[19] While different epistemological theories will give different explanations of this phenomenon, the idea that there's *something* to be accounted for here is relatively uncontroversial.[20]

[19] The example was introduced by Dretske (1970), though his interpretation of it is somewhat different—on his view, even while attending to the mule possibility, we can truly say that Susy knows that the animal is a zebra. We just cannot say that she knows that the animal is *not* a cleverly disguised mule.

[20] Perhaps the best-known strategy for explaining the phenomena is to opt for a version of epistemic contextualism, defended by Stine (1976), Cohen (1987), DeRose (1995), Lewis (1996), and many others. But there are myriad non-contextualist strategies for explaining this phenomenon too, e.g., sensitive invariantist views of the sort defended by Hawthorne (2004), Stanley (2005), and Fantl and McGrath

Slightly less familiarly, it's not just adding possibilities by "brute force"—"hey, what if it's a cleverly disguised mule?"—that may lead to our withdrawing knowledge ascriptions. Also, changing how we "carve up" a given set of possibilities may make new error possibilities salient, and thus affect our willingness to ascribe knowledge.[21] Consider a glass of water on a hot day with several ice cubes floating in it. Normally, we'll be happy to say that we know that the ice will melt—we'll be happy to simply ignore possibilities in which the water looks as it actually does, but in which the ice will get colder, rather than melting. But this way of dividing up the space of possibilities isn't inevitable. Rather than thinking of the relevant distinction in a binary way—the ice will melt, or it won't—we might instead decide that we should really be thinking of the situation in terms of all the possible microstates of the system consistent with its observable macrostate—i.e., all the specific ways the water molecules in the glass could be arranged such that what we end up seeing is a glass of water with a few ice cubes floating in it. And once we start thinking of things that way, at least if we know some statistical mechanics, it's very hard to ignore the fact that there is a set of microstates—admittedly a very tiny one—with the property that, if the system is in one of these microstates, then it will evolve such that the ice does not melt but instead gets colder. And once we do this, it's tempting to hold that, strictly speaking, we cannot know that the ice will melt. At best, we can know that it is extremely probable that it will melt.[22]

Examples of this sort can be multiplied. Not only can other areas of physics be called on to make the same philosophical point,[23] but more everyday examples, e.g., involving lotteries or similar setups, can work too. While I may initially be inclined to say that I know that my car is parked where I left it this morning, once I start thinking about the various possibilities in which different cars in my region are stolen, and think of possibilities in which *my* car is stolen as similar to those in which other cars are stolen, at least *some* of which are likely to obtain, I become much less inclined to say that I know where my car is parked.[24] More generally, just about whenever we're inclined to ascribe knowledge to some subject that P, there will be some very fine-grained way of thinking about the space of epistemic possibilities such that, when we think of things

(2009), epistemic relativism (MacFarlane, 2011), and even insensitive invariantist views on which the phenomena are accounted for in terms of the instability of belief (Nagel, 2010).

[21] Leitgeb (2014) offers a theory of belief on which whether a subject believes that P is "partition-dependent"—it depends, in a broadly contextualist way, on how possibilities are partitioned. While he doesn't discuss knowledge, there's no obvious reason his account couldn't be extended to do so.

[22] See Myrvold (2016).

[23] Hawthorne (2004) and Hawthorne and Lasonen-Aarnio (2009) appeal to examples from quantum mechanics to do similar work.

[24] See Vogel (1990).

that way, the knowledge ascription is much less attractive. It will look like a claim to the effect that we can know that a given ticket will lose a fair lottery.

Just what should we make of this? Some philosophers use such observations to motivate the idea that, strictly speaking, we have little if any knowledge.[25] This is strongly reminiscent of the idea that, strictly speaking, nothing is certain. While I think this isn't quite the right lesson to draw—again, see Chapter 4 for a similar dialectic concerning certainty—I won't revisit it here.

Rather, my point is that Lederman's argument against the possibility of groups attaining common knowledge should be thought of as belonging to the same family. That is, common knowledge is a "companion in guilt" with individual knowledge, and if phenomena like lotteries and statistical mechanics don't lead us to reject the possibility of individual knowledge, we should think hard before rejecting the possibility of common knowledge on the basis of cases like SAILBOAT.

In the remainder of this chapter I'll first introduce a framework for thinking about knowledge (including common knowledge) that has been influential in philosophy and cognate disciplines, especially in the context of discussions of common knowledge. I'll then use this framework to illustrate the phenomenon just alluded to; I'll show how, depending on a modeling choice about whether to represent possibilities in a coarse-grained or fine-grained way, certain knowledge ascriptions can seem more or less appropriate.[26] Finally, I'll turn to the question of comparing coarse-grained models of cases like SAILBOAT, in which common knowledge is attainable, to fine-grained ones in which it's not.

7.2.2 Common Knowledge in Distributed Systems

The framework I'll use in this chapter is one that computer scientists have found helpful in theorizing about "distributed systems."[27] I use this framework because (a) much theorizing about common knowledge has been done in this framework, and (b) the framework is consistent with Lederman's assumptions, so I don't beg any questions against him by using it. Abstractly, a distributed system involves various different individuals (perhaps computers, or wires, or people) embedded in an environment. The individuals and the environment

[25] See Hawthorne (2004, ch. 3) for some discussion without endorsement of the options available to a skeptic who uses these sorts of considerations to motivate her position.

[26] To be clear, even a "fine-grained" way of carving up possibilities here will be "coarse-grained" in the sense discussed in Chapter 3.

[27] See Fagin et al. (2003) for detailed discussion of applications in computer science.

Table 7.1 No Common Knowledge

<table>
<tr><th>Nature</th><th>Bob</th><th>Alice</th></tr>
<tr><td rowspan="2">Rain</td><td rowspan="2">Bob sees rain</td><td>Alice is behind Bob during rain</td></tr>
<tr><td rowspan="2">Alice is in other room</td></tr>
<tr><td rowspan="2">No Rain</td><td rowspan="2">Bob sees no rain</td></tr>
<tr><td>Alice is behind Bob during no rain</td></tr>
</table>

can each be in a variety of different local states, and the global state of the system is determined by the local states of the individuals and the environment that compose it. Crucially for my purposes, the framework allows that an individual knows that ϕ (where ϕ might be some fact about the environment, or about her fellow individuals) just when she's in a local state that she would only be in if ϕ. This framework can seem outlandishly idealized, in some ways that should be familiar—it uses possibilities to represent information, so it inherits the problems of coarse grain discussed in Chapter 3, and it represents agents as having perfect knowledge of their *own* epistemic states, so it goes in for exactly the sorts of level "conflations" discussed in the previous chapter. But Lederman is willing to grant these assumptions for the sake of argument; after all, the idea that obstacles to *self*-knowledge can create obstacles to common knowledge is already familiar in the philosophical literature—see Williamson (2000) and Hawthorne and Magidor (2009).

To see how this framework lets us reason about knowledge—especially knowledge of what others know—it will help to consider some examples. Let's start simple. Consider a system involving nature, Alice, and Bob. Nature has just two possible states—it can rain, or it can fail to rain. Bob is looking out the window, so he knows whether it's raining. Alice is standing behind Bob, also looking out the window. She also knows whether it is raining, *and* she knows that Bob knows that it's raining. But Bob can't see that Alice is behind him—for all Bob knows, Alice is in the other (windowless) room, and neither knows that it's raining, nor knows that Bob knows that it's raining. We can capture all this in a relatively simple diagram, labeled Table 7.1 above.

The actual global state of the system is represented by the top row—the one in which it's raining, Bob sees that it's raining, and Alice is behind Bob. The diagram represents Bob as knowing that it is raining in this situation because his local state—that of seeing rain—is one that he's only in when it is raining. The diagram also represents Alice as both knowing that it's raining and knowing that Bob knows this because her local state—that of being behind Bob during rain—is one that she's only in when it's raining and Bob knows that

Table 7.2 Common Knowledge

<table>
<tr><th>Nature</th><th>Bob</th><th>Alice</th></tr>
<tr><td rowspan="2">Rain</td><td>Bob is beside Alice during rain</td><td>Alice is beside Bob during rain</td></tr>
<tr><td>Bob sees rain</td><td rowspan="2">Alice is in other room</td></tr>
<tr><td rowspan="2">No Rain</td><td>Bob sees no rain</td></tr>
<tr><td>Bob is beside Alice during no rain</td><td>Alice is beside Bob during no rain</td></tr>
</table>

it is raining. But the diagram represents Bob as failing to know that Alice knows these things because his local state is compatible with her knowing neither of these things; the diagram represents it as possible that Bob sees rain, while Alice is in the other room. And when Alice is in the other room, she does not know that it is raining—this is represented by there being both rain and no rain possibilities that are compatible with Alice being in the other room. Moreover, the diagram represents Alice as knowing that Bob fails to know whether Alice knows whether it is raining—this is because Alice's state is one that she's only in when Bob fails to know what Alice knows.

The example just discussed didn't involve any common knowledge (or at least, none of a non-trivial sort), but the framework can easily allow for it. Let's consider a slight modification of our example. Suppose that when Alice is in the same room as Bob, rather than looking over his shoulder, she stands right next to him. This way, they can clearly see each other, in addition to being able to see through the window. This is represented in Table 7.2.

Let's consider what's true in the top row of the diagram—in which it's raining, and Bob and Alice can see the rain and each other. Both of them know that it is raining—each of them is in a local state that he/she is only in when it is raining. They also both know that they both know that it is raining—each is in a state that he/she is only in when they both know that it is raining. They also both know that they both know that they both know that it is raining, for the very same reason. It's easy to see that, according to the diagram, it is common knowledge between them that it is raining—for any numbers of iterations of "they both know that," they both know that they both know... that it is raining. By the same token, when they are standing beside each other and it is *not* raining, it is common knowledge between them that it is not raining.

I've just shown how the framework I've been discussing makes it straightforward to represent situations in which groups have common knowledge. Under certain circumstances in which everybody in a group of agents knows some fact, this framework allows that it will also be common knowledge among the group.[28]

Before discussing Lederman's argument, I'll digress to note that the present framework provides some support for the ideas about complexity discussed in §6.2.1. Distributed systems models in which agents have common knowledge needn't be particularly complex. Table 7.2 was simple enough, and could've been further simplified while still allowing for non-trivial common knowledge.[29] But as n increases, the complexity of the model we need in order to represent a situation in which agents have n but not $n + 1$ iterations of shared knowledge—as measured by the number of distinct states of the agents the model must contain—increases monotonically.

Table 7.1 above is a simple model—Bob has just two possible states—in which, when Alice is behind Bob during rain, Alice and Bob both know that it's raining, but they don't both know that they both know this (since Bob doesn't know that Alice knows it). But if we want to represent a situation in which Bob and Alice both know that it is raining, and both know that they both know this, *without* both knowing that they both know that they both know this, no model where either agent has just two states will do.

The top row of Table 7.3 represents a possible situation in which Alice and Bob have two but not three iterations of shared knowledge; they both know that they both know that it is raining, but they do *not* both know that they both know that they both know that it is raining. I explain in a footnote.[30] Hopefully once the reader has encountered Table 7.3, it should be clear enough how one would go about constructing a model representing three but not four iterations of shared knowledge, or four but not five, and also clear enough that

[28] General characterizations of the conditions under which this framework makes common knowledge possible can be found in Fagin et al. (2003).

[29] Just imagine a variant of table two in which the middle two rows are eliminated—either Alice is beside Bob during rain and its common knowledge that it's raining, or Alice is beside Bob during no rain, and it's common knowledge that it's not raining.

[30] Consider the top row of the model, where Bob is in B1 and Alice is in A1. Both of those states entail that it is raining (R), so we have:

$$K_B(R)\&K_A(R).$$

Also, B1 entails A1, which entails R, so Bob knows that Alice knows that it is raining. And of course, Bob knows that Bob knows that it is raining. And A1 entails B1 or B2, both of which entail R, so Alice knows that Bob knows that it is raining. And of course, Alice knows that Alice knows that it is raining. So we have our first iteration of shared knowledge:

$$K_B((K_B(R)\&K_A(R))\&K_A((K_B(R)\&K_A(R)).$$

I'll leave to the reader to verify the second iteration of shared knowledge. But it's also easy to see that they don't have common knowledge. A1 is consistent with B2, which is consistent with A2, which is consistent with B3, which is consistent with no rain. So Alice does *not* know that Bob knows that Alice knows that Bob knows that it's raining.

Table 7.3 Two but Not Three Iterations of Shared Knowledge

<table>
<tr><th>Nature</th><th>Bob</th><th>Alice</th></tr>
<tr><td rowspan="4">Rain</td><td>B1</td><td rowspan="2">A1</td></tr>
<tr><td rowspan="2">B2</td></tr>
<tr><td rowspan="2">A2</td></tr>
<tr><td rowspan="2">B3</td></tr>
<tr><td rowspan="2">No Rain</td><td rowspan="2">A3</td></tr>
<tr><td>B4</td></tr>
</table>

as we keep going, our models will need more and more possible states of the agents to represent such situations.

So the framework captures the intuition we got at in §6.2.1—that infinite iteration of the sort posited by the KK principle needn't mean intractable complexity, and can instead be simpler than finite iterations—via a different route: simple agents who make few distinctions can have common knowledge, but only more sophisticated agents making subtler distinctions can have five but not six iterations of shared knowledge.

But enough about complexity. The point that will be crucial in my discussion of Lederman's argument is that all these claims about who knows what, and what is common knowledge, are highly sensitive to which possibilities are included in the model. For instance, the choice in Table 7.1 to include just a single state that Bob can be in when nature is in the "Rain" state amounts to setting aside various skeptical possibilities. But the framework doesn't force us to do that. If we want to use the framework to illustrate the skeptical idea that Bob *cannot* know that it is raining by looking out a window, because he cannot rule out, e.g., that he is hallucinating, we could do that by labeling the states of Bob as mere "seemings," and representing him as being capable of being in the same such state when it rains as when it doesn't.

We can also use the framework to illustrate how our willingness to make knowledge ascriptions can be sensitive to the granularity with which possibilities are represented. Return to the example of a glass of ice water on a hot day, observed by an ordinary adult (again, call him Bob). We might represent a simple way of thinking about the situation using the framework in Table 7.4.

This captures the natural thought that, when Bob sees a typical glass of ice water, there's only one way for things to go vis-á-vis the ice melting: it will melt. But in Table 7.5 we represent the result of thinking of the situation in statistical mechanical terms, where microstates $1 - n$ are all the microstates

Table 7.4 No Skepticism

Nature	Bob
Ice will melt	Bob sees a typical glass of ice water

Table 7.5 Skepticism

<table>
<tr><th>Nature</th><th>Bob</th></tr>
<tr><td>Ice water system is in microstate 1</td><td rowspan="5">Bob sees a typical glass of ice water</td></tr>
<tr><td>Ice water system is in microstate 2</td></tr>
<tr><td>Ice water system is in microstate 3</td></tr>
<tr><td>⋮</td></tr>
<tr><td>Ice water system is in microstate n</td></tr>
</table>

that correspond to the observable macrostate of a typical glass of ice water, some small portion of which will evolve such that the ice doesn't melt.

This way of representing things captures the idea that Bob does not know that the ice will melt, since things would look just the same to Bob even if the system turned out to be in one of the microstates that leads to non-melting. And it brings out the fact that Table 7.4 leaves possibilities out—possibilities where Bob sees a typical glass of water, but the ice doesn't melt. Ignoring these possibilities may seem perfectly natural when we're thinking about the system using coarse-grained distinctions (will the ice melt?) but not so much when we're using more fine-grained distinctions (which microstate is the system in?).

7.2.3 Sailboat Revisited

Now let's return to Sailboat. Imagine yourself in the position of Roman or Columba, before the mast is shown. The following strikes me as a natural bit of planning to engage in:

> 100 cm is about 40 inches, a typical height for young child. If the mast we're shown is much greater than that—e.g., comparable to or taller than either of us, who are adults—then that will be clear to both of us, and I'll press the button, assuming my partner will do the same. If the mast we're shown is

Table 7.6 Coarse-Grained Sailboat

<table>
<tr><th>Sailboat</th><th>Alice</th><th>Bob</th></tr>
<tr><td>Mast is >>100 cm</td><td>Confident mast is >100 cm</td><td>Confident mast is >100 cm</td></tr>
<tr><td>Mast is >100 cm but not >>100 cm</td><td rowspan="2">Uncertain as to mast height</td><td rowspan="2">Uncertain as to mast height</td></tr>
<tr><td>Mast is <100 cm but not <<100 cm</td></tr>
<tr><td>Mast is <<100 cm</td><td>Confident mast is <100 cm</td><td>Confident mast is <100 cm</td></tr>
</table>

> much less than that—e.g., comparable to or less than the height of a cat—then that will be clear too, and I'll refrain from pressing, again assuming my partner will do the same. If it's somewhere in the middle—obviously shorter than an adult, obviously taller than a cat—then I'm not yet sure what I'll do, and it will depend on just how tall it looks to be, and on what it looks like my partner thinks. But I don't yet need to come up with a plan for those cases, since the right course of action might be obvious.

If this is how both Roman and Columba are thinking about the situation, I suggest the following would be a natural way to model their knowledge, with the actual situation, where the mast is 300 cm—over 9 feet, much taller than any adult—represented on the top row of Table 7.6.

This would be natural, I claim, in part because the coarse-grained, rough distinctions in the model correspond to the coarse-grained, rough distinctions that the agents themselves might use—at least, if you agreed that the imagined bit of internal monologue sounded like a natural one—in making their plans about what to do. And according to this model, in the top row it is common knowledge between Alice and Bob that the mast is taller than 100 cm. The explanation of why this is so is identical to the explanation of why, according to the model in Figure 2, where Alice is standing beside Bob during rain, it's common knowledge between them that it's raining. That is, both Alice and Bob are in local states that they're only in when the mast is taller than 100 cm—i.e., they both know the mast is taller than 100 cm—and they're both in local states that they're only in when they're both in local states that they're only in when the mast is taller than 100 cm—i.e., they both know that they both know that the mast is taller than 100 cm—and so on.

In the case of "lottery-style" skepticism, moving to a fine-grained way of thinking of things makes certain error possibilities—e.g., ones in which the ice-water system is in a thermodynamically atypical microstate, or ones in which *my* car is one among the few that will be stolen in my region this year—newly salient, and thereby leads us to withdraw knowledge ascriptions. I've suggested that we can think of Lederman's argument in the same way, but it may not be clear just how the analogy works—which are the *error* possibilities that we introduce when we move from the coarse-grained way of thinking of the mast's height to the fine-grained one?

If you look at Table 7.6, you'll notice it doesn't include any possibilities where Alice and Bob are in different coarse-grained states of confidence. When Alice is confident the mast is taller than 100 cm, so is Bob, according to Table 7.6. And likewise when she is uncertain, and when she is confident the mast is shorter than 100 cm. Plausibly, this assumption will be true for most heights the mast might be. But not all—there will be intermediate, borderline cases, to which Alice and Bob might respond differently. Suppose, e.g., that the mast is 137 cm—about four and a half feet. That's certainly less than a typical adult, but is it *so* clearly less that Alice and Bob will both be sure that the other classifies it in the same way? Mightn't someone very cautious count it as a borderline case or at least think that someone else might? (or that someone else might think that someone else might?) It's only because it leaves out these borderline cases that the model in Table 7.6 counts it as common knowledge that the mast height is >100 cm when it is in fact >>100 cm.

Perhaps this makes the verdict of the model seem obviously wrong, or simplistic. If it only classifies something as common knowledge due to *oversight*, then how can such a classification be of philosophical interest? In my opinion, this move would be far too quick. Rayo (2008) defends an attractive view on which vague concepts are only useful in thought and talk in "gappy" contexts—i.e., contexts where we are ignoring borderline cases for their application. The present position can be characterized as the idea that common knowledge will only be felicitously ascribable in similarly "gappy" contexts. E.g., suppose Rayo is right that we can only felicitously describe a person as "tall" in a context where it's *not* the case that every height in inches from 50 to 100 is treated as a live possibility for her height; we must be ignoring or setting aside certain intermediate heights for it to be felicitous to use "tall" to characterize someone's height. What Rayo says is required for uses of "tall" to be felicitous is exactly what I'm suggesting is required for ascriptions of common knowledge that someone is tall to be felicitous.

Taking stock, what we've seen so far is that cases like SAILBOAT seem to allow for groups of agents to have common knowledge if, but only if, we think of them in relatively coarse-grained ways, which will involve ignoring certain possibilities. When we make much more fine-grained distinctions among possible states of nature and epistemic states of agents, and ignore fewer possibilities, common knowledge ascriptions are much less attractive.

It's tempting to think that making finer-grained distinctions among possible states of nature will make a model more accurate—more similar to the real-world target systems it represents—albeit perhaps at some cost in tractability. So if common knowledge seems to disappear when we make our models more discriminating, shouldn't we accept that models incorporating common knowledge are in some sense more idealized and less accurate than their counterparts that eschew it?

This is what both Lederman and Williamson suggest. As I interpret them, the strict truth is that our interpersonal knowledge always gives out after some specific number of iterations, but models in which it doesn't are sometimes handier to use than models in which it does. As Williamson (2000, p. 122) puts it, "Common knowledge would therefore be a convenient idealization, like a frictionless plane." Lederman (2018a) makes a similar analogy between common knowledge and the extensionless point masses of classical mechanics. I expressed some skepticism about this sort of analogy in the previous chapter, but now I'll go into a bit more depth.

In my view, phenomena in physics where we believe that there is an exact value for some parameter, but in which we model it as 0 (or infinity) for the sake of convenience are quite different from the strategic phenomena we model using common knowledge. The analogy would be a reasonable one if the sort of idealization involved in game theoretic models were "Galilean" idealization, in a sense I'm about to describe. But, ultimately, there's good reason to doubt that it is.

One of the most familiar justifications for idealization in science involves computational tractability; sometimes, we have a sense of what a highly accurate model of some system would look like, but the model would be too complex to work with, so we construct and analyze a related but simpler model instead. The simpler model is obtained by starting with the complex one, and then ignoring certain factors, or setting certain parameters to zero. McMullin (1985) and later Weisberg (2013) call this practice "Galilean idealization," after Galileo, who self-consciously employed such a practice in his theoretical and experimental work. Galilean idealization is, in principle, eliminable. The factors ignored, or the parameters set to zero—friction and air resistance, in Galileo's case—could in principle be added back in, albeit at high computational cost, and perhaps with little predictive or explanatory benefit.

If you think the only sort of idealization a domain of inquiry calls for is Galilean idealization, then your approach to model-building in that domain is not modest, in my sense. If you think all idealization in some domain is Galilean idealization, then you think it's eliminable; you'll think that Plato's heaven contains a non-idealized, maximally accurate model of the phenomena in the relevant domain, even if we humans are too computationally limited to identify and solve that model. As Weisberg (2013, p. 100) puts it, "Galilean idealization takes place with the expectation of future de-idealization and more accurate representation."

But not all idealization is Galilean idealization.[31] Colin Rice (2018, 2019) offers a number of nice examples of models as "holistically distorted" representations of the systems they model. In these cases, rather than starting with an accurate but intractable model that becomes tractable when we set some parameter to zero (or ∞), scientists build a model that involves numerous, radical distortions of its target. Moreover, these distortions don't merely involve making a complicated calculation simpler; rather, absent the distortions, the mathematical techniques that work for solving the model would simply be inapplicable.

Because I think his discussion is highly relevant to models of knowledge, I want to dwell a bit on his treatment of optimality models in evolutionary biology. Optimality models attempt to explain animal behavior by showing that it is an optimal response to the constraints and tradeoffs presented by some ecological niche. For example, he discusses a model offered by Heath et al. (2010) in which some prima facie puzzling features of duck foraging behavior are explained as an energetically optimal response to tradeoffs and constraints involving digestion, the tide cycle, and the times at which foraging is most profitable. In what sense is this model—which is typical of optimality models—holistically distorted? Rice (2018) explains:

> Indeed, biological optimality models typically make use of a large collection of idealizing assumptions, including (but not limited to):
>
> 1. The models' smooth mathematical curves and simplified (e.g., discrete) equations are idealized when compared to actual processes within the models' target system(s).

[31] Weisberg (2013) describes two other types of idealization besides Galilean, which he calls "minimalist idealization" and "multiple models idealization." The latter is closely related to my project.

2. The strategy sets are intended to capture the relevant set of alternatives rather than the set of strategies actually competing in the population's past.[32]
3. There are idealizations regarding the processes of inheritance; e.g., assuming that offspring will perfectly resemble their parents and that reproduction is asexual.
4. The models assume there is no intergenerational overlap.
5. The models' optimization assumptions (e.g., maximization of energy intake) do not accurately represent an actual selection process in the system, but only captures a general optimizing tendency of the system in the long run.
6. It is assumed that selection pressures in the population do not change over time.
7. The individual-level events that determine the fitnesses of the trait types are assumed to be random and statistically independent to allow the population to be represented as a normal distribution of trait types.
8. Infinite population size is assumed to allow for the use of various laws of large numbers that eliminate statistical error (i.e., drift).

He goes on to explain why this host of idealizations amounts to a holistic distortion of the target system, in a way that precludes separating out the contributions of the accurate and inaccurate parts of the model:

> While these idealizations result in a pervasive distortion of the actual evolutionary processes, making these assumptions enables biological modelers to apply the central limit theorem and various laws of large numbers in order to make exact calculations about the equilibrium point predicted by the optimization model. In other words, these idealizing assumptions are *necessary* for applying the mathematical techniques used in the explanation. As a result, it is impossible to isolate the contributions made by the accurate parts of these mathematical models from the contributions made by their idealized parts. Only by pervasively distorting the features of real-world systems can biologists apply the mathematical modeling techniques required to extract the modal information used to explain.

[32] As an aside, the choice of strategies to include in a model is plausibly a locus for modesty in model selection within the framework of optimality modeling. Just which strategies to include in a model, and at what level of grain to distinguish them, are matters for experience-informed judgment. Likewise for the question of which constraints and tradeoffs the model should represent.

Rice's purpose in discussing holistically distorted models is to argue against the view that we can understand successful, explanatory models as those whose inaccurate parts represent features of the target system that are irrelevant, or which make only small differences to the target phenomenon.[33] But the key lesson for the present discussion is that when models are holistically distorted, we can't always make them "more realistic" along some dimension of variation, and still have a functioning model of their target phenomenon. We've already seen that when a model involves Galilean idealization, there will be de-idealized versions of the model that are less tractable, but more accurate. But when a model is holistically distorted, there's no guarantee that de-idealizing it will involve this kind of tradeoff; it might simply break the model.

Consider an ecological model in which animal populations are represented by real numbers, and rates of change of those populations represented by differential equations.[34] Representing population sizes with real numbers is obviously an idealization; there can't be π rabbits.[35] But it might be an ineliminable one; if rates of growth and decay are represented by differential equations, then population size must vary continuously. Forcing populations to take discrete values would require a fundamentally different set of mathematical tools, and there's no a priori guarantee that a comparably powerful and explanatory model using such tools will exist; it might very well be that the patterns picked out by certain models in population ecology simply can't be identified using models that involve more realistic assumptions about population sizes.[36]

It's worth considering whether the same might be true of the sort of idealizations involved in constructing coarse-grained models (like that in Table 6) in decision and game theory. Basic principles in game theory like the iterated elimination of dominated strategies only work when there are dominated strategies to be eliminated. But any model that represents a strategy as dominated could be "enriched" by adding states—perhaps remote, improbable ones—where that strategy turns out for the best. In such a richer model, the strategy won't be eliminable by dominance reasoning, and so any predictions to the effect that agents won't pursue it will have to be generated via other means.

[33] That is, he's arguing against a particular answer to the question of when idealizations are "harmless," raised way back in Chapter 1.

[34] See, e.g., Weisberg (2013)'s discussion of the Lokta-Volterra model in population ecology.

[35] As Williamson (2017) observes.

[36] Of course, sometimes differential equations can be effectively replaced by difference equations, which take discrete inputs and give discrete outputs. But in general there's no guarantee to this effect.

As I interpret Lederman, he's aware of this, and thinks that such other means will always be available.[37] In particular, any predictions generated by applying equilibrium reasoning to coarse-grained, tractable models in which players have common knowledge of the setup could instead be generated—perhaps less elegantly—by decision theoretic reasoning. Players can be represented as having a prior probability distribution over the possible acts of the other parties, and can be predicted to play expected utility maximizing strategies, rather than game-theoretic equilibrium strategies. While I'm certain that this is *sometimes* a fruitful approach, I see no reason to assume that it will always be available. As we've just seen, sometimes holistically distorted models can't be made more realistic along some dimension without breaking. And the models of common knowledge Lederman is discussing involve a wide enough variety of idealizations—concerning logical omniscience (Chapter 3), certainty (Chapters 4 and 5), self-knowledge (Chapter 6), and a coarse-grained representation of the possible states of nature (this chapter)—that it seems reasonable to regard them as holistically distorted.[38] They're not like models in physics that could in principle be perfectly accurate except for some fudging about friction. So we shouldn't assume that whatever explanatory power they have could be replicated—albeit perhaps more clumsily—by models that are more realistic in the granularity with which they represent the states of nature.

In particular, Lederman's preferred route strikes me as dangerously unconstrained; it suggests an approach where predictions about which strategies agents will play effectively have to be "added in by hand" via the choice of the agents' prior expectations concerning how the other agents will behave. But it's a robust feature of economic practice that economists often prefer highly constrained models to superficially more realistic ones that are also more flexible.[39] If modeling agents as having common knowledge allows for constraints on which predictions the models can make—because common knowledge allows game theoretic reasoning to get some traction and rule out dominated strategies—that's an explanatory advantage that shouldn't be taken lightly. For all Lederman has shown, models without common knowledge wouldn't just be more awkward than models with it; they'd be less explanatorily powerful, because they would be flexible enough to accommodate any behavior you like.

[37] See especially the conclusion of Lederman (2018a).

[38] Note that even Lederman's preferred decision theoretic models would still be highly idealized in many of these respects. In particular, in modeling states of nature as differing only along one continuous dimension—e.g., the height of the mast of a sailboat—they leave out lots of detail.

[39] See, e.g., Morris (1995) for an illuminating discussion of the practice in economics of modeling agents as having common priors.

I don't take myself to have shown that a Ledermanian research program—one in which phenomena traditionally modeled using common knowledge are instead represented by models that vindicate interpersonal ignorance—is hopeless. Rather, my intended moral is a broadly Kuhnian one. *Until* such a research program has some substantial victories under its belt—until there's a range of phenomena that its models can elegantly explain, but which traditional common knowledge models can't—it's reasonable to continue defaulting to models incorporating common knowledge to explain a wide range of strategic interactions. While it's undeniably true that models incorporating common knowledge are idealized, it's far from obvious that broadly similar, partially de-idealized models—models involving agents, utility functions, information available to the agents, and some rules for predicting agents' behavior on the basis of the previous, but where the information available to the agents doesn't include any common knowledge—can do the same work. In the next and final chapter I'll generalize this Kuhnian lesson.

8

Ideal and Non-Ideal Epistemology

Political philosophers have long debated the proper relationship between so-called "ideal" and "non-ideal" political theory. The distinction is from Rawls, who understood the task of ideal theory as that of describing a perfectly just society, while non-ideal theory involved asking questions about how less than fully just societies might move in a marginally more just direction.[1] Recent years have increasingly seen epistemologists using similar ideal/non-ideal distinctions in thinking about their own subject matter, often to urge—as is common in much recent political philosophy[2]—that non-ideal theory has been unduly neglected. In this brief final chapter I'll examine whether and how we might draw such a distinction within the framework of the present book. I'll use Jennifer Carr's "Why Ideal Epistemology?" (2021) as a foil, for a couple of reasons. First, it's the most explicit and general presentation I'm familiar with of a view that, it seems to me, is widely shared among at least a subset of epistemologists. And second, while the framework it presents is a prima facie appealing one, I'll argue that it ultimately turns out to be much more fraught than might at first appear, in light of the perspective on modeling and idealization I've defended in this book.

As Carr draws the distinction, "Ideal epistemologists are concerned with questions about what perfectly rational, cognitively idealized, computationally unlimited believers would believe. . . . By contrast, nonideal epistemologists are concerned with questions about epistemic norms that are satisfiable by most humans much of the time." (p. 1132) Much of Carr's interest in drawing the distinction is to defend the program of ideal epistemology. Her approach is, in broad outlines, similar to Rawls' understanding of the role of ideal political theory. Just as the non-ideal political theorist may advocate for an unjust policy because it's the best among the politically feasible options, a non-ideal epistemologist may recommend an epistemically non-ideal belief state—one that involves uncertainty in some logical truth, for instance—because it's the best among the cognitively feasible options. Non-ideal epistemology, in Carr's

[1] Rawls (1971). [2] E.g., Mills (2005).

Idealization in Epistemology: A Modest Modeling Approach. Daniel Greco, Oxford University Press.
DOI: 10.1093/oso/9780198860556.003.0009

view, is what you get by starting with ideal epistemology and then adding in various "ought-implies-can" style constraints.

For instance, ideal epistemic agents have bodies of belief that are deductively closed, while our bodies of belief are deductively open. Ideal epistemic agents have credences that are probabilistically coherent, while our credences are probabilistically incoherent. Ideal epistemic agents are certain of a great many things—perhaps all a priori truths, as well as some contingent facts about their own epistemic and/or conscious states—while limited agents like us are uncertain about all these matters.[3] But in each case, we can best complete the project of non-ideal epistemology by starting with a picture of ideal belief states, and then inquiring into how the non-ideal belief states we *actually* have could be further from or closer to the ideal along various dimensions.

I hope the reader already has an inkling of the shape my concerns will take. While I agree that modeling us as having probabilistically coherent credences involves idealization, it doesn't follow that we *actually* have probabilistically *in*coherent credences. And while I agree that modeling us as being certain of our evidence in particular and of the limits of our knowledge more generally also involves idealization, it doesn't follow that we *actually* are *un*certain about what our evidence is and what we know. To use an imperfect analogy, we can agree that it would be inaccurate to say that the king of France is bald without thereby committing ourselves to his being hirsute. Why does this matter? Ultimately, I'll argue that the neat division between ideal and non-ideal theory is easiest when the idealizations eschewed by non-ideal theory are purely normative. When "ideal" theory involves both descriptive and normative idealizations—perhaps inextricably linked—eschewing them is harder, and pitfalls lurk for the unwary.

I've just alluded to a distinction between *descriptive* and *normative* idealization, also drawn by Carr. Both descriptive and normative idealizers build models with features that are not instantiated in the actual world. When we engage in descriptive idealization, the reason why we're OK with this mismatch is because models with such features are more tractable, or simpler, or have more explanatory power, or something along those lines; descriptive

[3] As mentioned, while Carr is the most explicit defender of this way of dividing up the terrain, it seems to me that many contemporary epistemologists see things similarly. David Chalmers (2012) holds that while we are uncertain about a wide range of a priori matters, a certain kind of idealized agent wouldn't be (see especially the fifth excursus). Declan Smithies (2019b) takes a similar position, holding that ideal agents are certain of all a priori truths, as well as facts about what their evidence is and facts about the extent of their justified beliefs. But non-ideal agents like us can't be doxastically justified in believing those claims that ideal agents are certain of, so our beliefs don't obey such constraints. John Pittard (2019) distinguishes between what an agent "ideally ought" to believe, and what she ought to believe "given her rational abilities and limitations" (p. 171).

idealization is what we've encountered in this book thus far. In normative idealization, the reason we're OK with mismatch between the model and the world is very different—the model represents how the world *should* be, or how it would be *good* for it to be.

While the distinction is simple enough conceptually, it's not always easy to apply, in large part because the very same features of the very same models are sometimes understood as descriptive, sometimes as normative. Both economists and philosophers construct and analyze models of expected-utility maximizing, logically omniscient, probabilistically coherent agents. Everybody will agree that such models are, in various senses, idealized. But in my experience at least, economists are more likely to think of the idealization in such models as purely descriptive; of course *we* are not perfect EU-maximizers, but models of such agents are simple and powerful, and provide a fruitful explanatory benchmark for making sense of actual human behavior. Ask economists whether the EU-maximizing agent represents a normative ideal, and they'll usually demur, insisting that, in their capacity as economists, they make no normative claims.[4] By contrast, philosophers tend to see the logically omniscient EU-maximizing agent as a kind of normative ideal—it represents a person who makes no errors of a certain kind—but if you ask them whether such models are predictively useful *they'll* demur and insist that in their capacity as philosophers they make no *empirical* claims.

Why does this matter? Because eschewing purely normative idealizations is typically easier than eschewing idealizations that play a descriptive role. Imagine a non-ideal political theorist interested in wealth inequality. She thinks of the ideal with respect to equality as a society where everyone has the same amount of wealth. But that leaves lots of room for questions about how to rank non-ideal societies with respect to the value of equality. She might argue that some measure of inequality—perhaps some version of the Gini coefficient[5]—gives us the right picture not only of the ideal (a Gini coefficient of 0), but also of what amounts to getting closer to it; the lower the Gini coefficient, the better.

For such a theorist, a Gini coefficient of 0 would be functioning as a purely normative ideal. Why purely normative? Because a model of a society with a Gini coefficient of 0 is no simpler, or more explanatorily powerful, or scientifically fruitful than a model of a society with a positive Gini coefficient.

[4] Milton Friedman's "The Methodology of Positive Economics" (1953) is probably the locus classicus for the view that economists qua economists should refrain from making normative claims.

[5] See Eliazar (2018) for a recent survey of inequality indices.

The frameworks economists typically use for representing wealth distributions can just as smoothly represent unequal distributions as they can represent equal ones; it's not as if adding in inequality plays a role analogous to adding in friction or air resistance in models of mechanics.

Some examples of "non-ideal" epistemology are similar. Back in Chapter 1 I alluded to agent-based models of epistemic communities. Using the same sorts of familiar descriptive idealizations we've been discussing throughout this book—probabilistic coherence, Bayesian updating, and the like—modelers can distinguish "ideal" situations where relations of trust allow reliable information to quickly spread throughout an epistemic community, and "non-ideal" ones where information remains siloed, or where false information easily spreads.[6] And such models allow for a project along the lines Carr discusses, wherein we ask how non-ideal communities might move in directions that make such failure modes less likely.[7] Here too, the sense in which the community where reliable information flows freely is "ideal" is a purely normative one; agent-based modeling frameworks have just as easy a time representing communities with "non-ideal" trust relations as with "ideal" ones; the computations are no less tractable in the one case than the other, nor are there conceptual problems with understanding or interpreting models of non-ideal epistemic communities.

But in cases where the ideals of ideal epistemology are descriptive—or inextricably descriptive and normative, as coherence ideals plausibly are—the project of non-ideal epistemology is much trickier. This is because, at least in practice, descriptive idealization is much more likely to involve the kind of "holistic distortions" we discussed in the previous chapter, and so is much more likely to be resistant to "de-idealizing." As a proof of concept, let's start with the ideal of transitivity of preference. My impression is that philosophers tend to see this as a purely normative ideal; the received view is that in fact, we often have intransitive preferences, but insofar as we do this represents a mistake. By contrast, economists—at least in their textbooks—tend to present things quite differently. There, it's common to use the "revealed preference" framework, in which an agent only has preferences at *all* if her choices obey certain constraints. In the revealed preference framework, facts about choice are treated as data which can, if they obey certain structural constraints, be

[6] See, e.g., Singer et al. (2019) and O'Connor and Weatherall (2019).

[7] Though see also Zollman (2007), who presents models in which limits to communication can be epistemically beneficial. In a nutshell, communities where information flows a bit less freely can avoid jumping to conclusions too quickly.

summarized in terms of a preference ordering.[8] In this framework, transitivity of preference isn't naturally thought of as a normative ideal, since it isn't possible to violate. Any set of choices that obeys the revealed preference axioms gives rise to a set of transitive preferences. For a set of choices that does *not* obey the revealed preference axioms, there's no set of preferences it gives rise to. The framework doesn't give us a recipe for generating an intransitive set of preferences from some set of choices.

While the revealed preference framework probably strikes most contemporary philosophers as retrograde—unreconstructedly behaviorist—I hope the reader will bear with me as we explore how certain "non-ideal" projects look from the perspective of a partisan of that framework; ultimately, I think there are general lessons here, even if we don't agree that talk of preference is merely a way of summarizing facts about choice.

Suppose, thinking of transitivity of preference as a purely normative ideal, some non-ideal decision theorist decides to ask how someone with non-transitive preferences ought to behave, and/or how such a person could approach the ideal of transitivity even without getting all the way there. These questions should look ill-posed to our revealed preference theorist, at least until she hears more about some generalization of the revealed preference framework the decision theorist might have in mind. Perhaps with some weaker constraints, our decision theorist might be able to generate a set of (possibly intransitive) preferences from a wider range of patterns of choice than those that can be represented in standard revealed preference theory. Perhaps not. But that's the kind of thing our revealed preference theorist would need to see before she'd be comfortable asking the sorts of questions our non-ideal decision theorist wanted to start with. Until such a generalization is provided, our revealed preference theorist could rightfully insist that she doesn't really know what we're talking about when we ask how someone with intransitive preferences ought to behave. She regards preference talk as a handy framework for modeling facts about choice, and she doesn't (yet) know what facts about choice would be fruitfully modeled by a set of intransitive preferences.

Similar issues arise when non-ideal epistemologists ask how someone with inconsistent beliefs should reason,[9] or which ways of being probabilistically

[8] E.g., the following gives a rough idea (but is too strong, in that it rules out indifference): if an agent sometimes picks A when B is available, she'll never pick B when A is available. See (Rubinstein, 2005, chs. 3–4) for a typical exposition of the revealed preference paradigm.

[9] E.g., Priest (2001), though he's perhaps not rightly thought of as engaging in "non-ideal" epistemology here, in that he does not think of inconsistent beliefs as necessarily being non-ideal in any sense.

incoherent are closer to the ideal of coherence than others,[10] or how someone who doesn't know what her evidence is should update her beliefs.[11] To the extent that we understand talk of beliefs, credences, and evidence as part of a modeling framework for holistically making sense of behavior—a perspective I defended in Chapter 3—then we should only be comfortable with these questions to the extent that we have a sense of what sort of behavior would be fruitfully modeled with inconsistent beliefs, or probabilistically incoherent credences, or evidence subjects don't know they have. And in each case, I've tried to show in previous chapters that these are thornier questions than they might initially seem.

None of this is meant as an argument to the effect that the "non-ideal" epistemological projects just mentioned can't be fruitfully pursued. But it is meant to show that they have more work to do than one might initially realize. When "non-ideal" theory involves relaxing purely normative idealizations, it's simpler. While questions about how to rank non-ideal states may be thorny, as well as questions about how to feasibly transition from lower to higher-ranked states, no special problems arise concerning how to describe or interpret non-ideal models. By contrast, when non-ideal theory involves relaxing the sorts of holistic, (at least partly) descriptive idealizations I've discussed in this book, the thorny questions about ranking and transitions only arise *after* some more fundamental conceptual questions are resolved. If we don't know what sorts of concrete situations are represented by "non-ideal" models, we're not yet in a position to ask evaluative questions about how they should be ranked against each other or normative questions about which non-ideal states we should aim for.

Ultimately, this conclusion shouldn't come as too much of a surprise. Throughout this book I've argued that idealization is ubiquitous and inescapable in epistemology. As such, there isn't a feasible, general project of "de-idealizing" the field. While particular models can fruitfully be made more realistic or detailed along this or that dimension, such projects always need to be defended one at a time, on a case-by-case basis. Carr's piece is written from the perspective of a friend of "ideal theory" in epistemology, eager to persuade skeptics of its value. The perspective suggested by the present book is much more unapologetic. We're all ideal theorists, and inevitably so.

[10] See Staffel (2019). [11] See Gallow (2021).

Bibliography

Alquier, Pierre. 2020. "Approximate Bayesian Inference." *Entropy* 22(11):1272.

Alston, William P. 1980. "Level Confusions in Epistemology." *Midwest Studies in Philosophy* 5(1):135–150.

Ando, Tomohiro. 2010. *Bayesian Model Selection and Statistical Modeling*. CRC Press.

Appiah, Kwame Anthony. 2017. *As If: Idealization and Ideals*. Harvard University Press.

Aumann, Robert. 1976. "Agreeing to Disagree." *Annals of Statistics* 4:1236–1239.

Bach, Kent. 2008. "Applying Pragmatics to Epistemology." *Philosophical Issues* 18:68–88.

Baddeley, Michelle. 2014. "Rethinking the Micro-foundations of Macroeconomics: Insights from Behavioral Economics." *European Journal of Economics and Economic Policies: Intervention* 11(1):99–112.

Batterman, Robert W. 2009. "Idealization and Modeling." *Synthese* 169:427–446.

Beaver, David, and Jason Stanley. 2019. "Towards a Non-Ideal Philosophy of Language." *New School Graduate Faculty Philosophy Journal* 39(2):501–545.

Beddor, Bob. 2020. "Certainty in Action." *Philosophical Quarterly* 70(281):711–737.

Beddor, Bob. forthcoming. "New Work for Certainty." *Philosophers' Imprint*.

Bergmann, Michael. 2006. *Justification without Awareness*. Oxford University Press.

Berto, Francisco, and Mark Jago. 2018. "Impossible Worlds." In "Stanford Enyclopedia of Philosophy," https://plato.stanford.edu/entries/impossible-worlds/.

Bicchieri, Christina. 1989. "Self-Refuting Theories of Strategic Interaction: A Paradox of Common Knowledge." *Erkenntnis* 30:69–85.

Blackburn, Simon. 1993. *Essays in Quasi-Realism*. Oxford University Press.

Bramson, Aaron, Patrick Grim, Daniel J. Singer, William J. Berger, Graham Sack, Steven Fisher, Carissa Flocken, and Bennett Holman. 2017. "Understanding Polarization: Meanings, Measures, and Model Evaluation." *Philosophy of Science* 84:115–159.

Briggs, R. A. 2009. "Distorted Reflection." *Philosophical Review* 118:59–85.

Bright, Liam Kofi. 2017. "On Fraud." *Philosophical Studies* 174:291–310.

Bronfman, Aaron. 2014. "Conditionalization and Not Knowing That One Knows." *Erkenntnis* 79:871–892.

Brueckner, Anthony. 2011. "Debasing Scepticism." *Analysis* 71:295–297.

Buchak, Lara. 2013a. "Belief, Credence, and Norms." *Philosophical Studies* 169(2):1–27.

Buchak, Lara. 2013b. *Risk and Rationality*. Oxford University Press.

Burge, Tyler. 1991. "Individualism and the Mental." In "The Nature of Mind", ed. David M. Rosenthal, pp. 536–568. Oxford University Press.

Burian, Richard. 1985. "On Conceptual Change in Biology: The Case of the Gene." In "*Evolution at a Crossroads. The New Biology and the New Philosophy of Science*", is a book published by MIT press.

Burian, Richard. 2004. "Molecular Epigenesis, Molecular Pleiotropy, and Molecular Gene Definitions." *History and Philosophy of the Life Sciences* 26:59–80.

Calabresi, Guido. 2003. "The Exclusionary Rule." *Harvard Journal of Law and Public Policy*. 26:111.

Callender, Craig. 1999. "Reducing Thermodynamics to Statistical Mechanics: The Case of Entropy." *Journal of Philosophy* 96:348–373.

Carnap, Rudolf. 1950. *Logical Foundations of Probability*. University of Chicago Press.

Carnap, Rudolf. 1967. *The Logical Structure of the World*. University of California Press.

Carr, Jennifer. 2021. "Why Ideal Epistemology?" *Mind*. 131(524), October 2022, pp. 1131–1162.

Carroll, Sean. 2016. *The Big Picture: On the Origins of Life, Meaning, and the Universe Itself*. Dutton Books.

Cartwright, Nancy. 1983. *How the Laws of Physics Lie*. Oxford University Press.

Cartwright, Nancy. 1999. *The Dappled World: A Study of the Boundaries of Science*. Cambridge University Press.

Chalmers, David. 2012. *Constructing the World*. Oxford University Press.

Chalmers, David. 2013. "Panpsychism and Panprotopsychism." *Amherst Lecture in Philosophy* 8.

Chalmers, David J. 1996. *The Conscious Mind*. Oxford University Press.

Chalmers, David J. 2011a. "Frege's Puzzle and the Objects of Credence." *Mind* 120: 587–635.

Chalmers, David J. 2011b. "The Nature of Epistemic Space." In *Epistemic Modality*, ed. Andy Egan and Brian Weatherson. Oxford University Press.

Chisholm, Roderick M. 1973. *The Problem of the Criterion*. Marquette University Press.

Chomsky, Noam. 1965. *Aspects of the Theory of Syntax*. MIT Press.

Christensen, David. 1992. "Confirmational Holism and Bayesian Epistemology." *Philosophy of Science* 59:540–557.

Christensen, David. 2004. *Putting Logic in its Place*. Oxford University Press.

Churchland, Patricia Smith. 1983. "Dennett's Instrumentalism: A Frog at the Bottom of the Mug." *Behavioral and Brain Sciences* 6(3): 358–359.

Churchland, Patricia. 1986. *Neurophilosophy: Towards a Unifiied Science of the Mind/Brain*. MIT Press.

Clark, Andy. 2016. *Surfing Uncertainty: Prediction, Action, and the Embodied Mind*. Oxford University Press.

Clark, Herbert, and Catherine Marshall. 1981. "Definite Reference and Mutual Knowledge." In *Elements of Discourse Understanding*, ed. Aravind Joshi, Bonnie Webber, and Ivan Sag, pp. 10–63. Cambridge University Press.

Clarke, Roger. 2013. "Belief Is Credence One (in Context)." *Philosophers' Imprint* 13:1–18.

Cohen, Laurence Jonathan. 1977. *The Probable and the Provable*. Clarendon Press.

Cohen, Stewart. 1987. "Knowledge, Context, and Social Standards." *Synthese* 73:3–26.

Cohen, Stewart. 1999. "Contextualism, Skepticism, and the Structure of Reasons." *Nous* 33:57–89.

Colyvan, Mark. 2013. "Idealisations in Normative Models." *Synthese* 190:1337–1350.

Cooper, Gregory F. 1990. "The Computational Complexity of Probabilistic Inference Using Bayesian Belief Networks." *Artificial Intelligence* 42:393–405.

Crawford, Vincent P., Miguel A. Costa-Gomes, and Nagore Iriberri. 2013. "Structural Models of Nonequilibrium Strategic Thinking: Theory, Evidence, and Applications." *Journal of Economic Literature* 51:5–62.

Das, Nilanjan. 2019. "Accuracy and Ur-Prior Conditionalization." *Review of Symbolic Logic* 12:62–96.

Das, Nilanjan. 2020. "The Value of Biased Information." *British Journal for the Philosophy of Science* https://doi.org/10.1093/bjps/axaa003.

Das, Nilanjan. 2022. "Externalism and Exploitability." *Philosophy and Phenomenological Research* 104:101–128.

Davidson, Donald. 1973. "Radical Interpretation." *Dialectica* 27:314–328.

Dennett, Daniel C. 1968. *Content and Consciousness*. Routledge.

Dennett, Daniel. 1981a. "Brain Writing and Mind Reading." In "*Brainstorms*," Bradford Books. pp. 43–56.

Dennett, Daniel. 1981b. "True Believers: The Intentional Strategy and Why It Works." In *Scientific Explanation: Papers Based on Herbert Spencer Lectures Given in the University of Oxford*, ed. A. F. Heath, pp. 150–167. Clarendon Press.

Dennett, Daniel. 1987. *The Intentional Stance*. MIT Press.

Dennett, Daniel. 1988. "Quining Qualia." In Consciousness in Contemporary Science, ed. A. Marcel and E. Bisiach, pp. 42–77. Oxford: Oxford University Press.

Dennett, Daniel. 1991. "Real Patterns." *Journal of Philosophy* 88:27–51.

DeRose, Keith. 1995. "Solving the Skeptical Problem." *Philosophical Review* 104:1–52.

DeRose, Keith. 2005. "The Ordinary Language Basis for Contextualism, and the New Invariantism." *Philosophical Quarterly* 55:172–198.

Dorr, Cian, Jeremy Goodman, and John Hawthorne. 2014. "Knowing against the Odds." *Philosophical Studies* 170:277–287.

Dorst, Kevin. Unpublished Manuscript. "Rational Polarization."

Dretske, Fred I. 1970. "Epistemic Operators." *Journal of Philosophy* 67:1007–1023.

Dupré, John. 1993. *The Disorder of Things: Metaphysical Foundations of the Disunity of Science*. Harvard University Press.

Elga, Adam, and Agustin Rayo. 2021. "Fragmentation and Information Access." In *The Fragmented Mind*, ed. Cristina Borgoni, Dirk Kindermann, and Andrea Onofri, pp. 37–53. Oxford University Press.

Elgin, Catherine. 2007. "Understanding and the Facts." *Philosophical Studies* 132:33–42.

Elgin, Catherine Z. 2009. "Is Understanding Factive?" In *Epistemic Value*, ed. Adrian Haddock, Alan Millar, and Duncan Pritchard, pp. 322–330. Oxford: Oxford University Press.

Eliazar, Iddio. 2018. "A Tour of Inequality." *Annals of Physics* 389:306–332.

Enoch, David, Levi Spectre, and Talia Fisher. 2012. "Statistical Evidence, Sensitivity, and the Legal Value of Knowledge." *Philosophy and Public Affairs* 40:197–224.

Fagin, Ronald, Joseph Halpern, Yoram Moses, and Moshe Vardi. 2003. *Reasoning about Knowledge*. MIT Press.

Fantl, Jeremy, and Matthew McGrath. 2009. *Knowledge in an Uncertain World*. Oxford University Press.

Field, Hartry. 1978. "Mental Representation." *Erkenntnis* 13:9–61.

Fitelson, Branden, and James Hawthorne. 2010. "How Bayesian Confirmation Theory Handles the Paradox of the Ravens." In *The Place of Probability in Science*, ed. Ellery Eells and James Fetzer, pp. 247–275. Springer.

Fodor, Jerry A., and Ernest Lepore. 1992. *Holism: A Shopper's Guide*. Blackwell.

Franklin-Hall, Laura. 2007. "Bacteria, Sex, and Systematics." *Philosophy of Science* 74:69–95.

Friedman, Michael. 1974. "Explanation and Scientific Understanding." *Journal of Philosophy* 71:5–19.

Friedman, Milton. 1953. *Essays in Positive Economics*. University of Chicago Press.

Gallow, J. Dmitri. 2021. "Updating for Externalists." *Noûs* 55:487–516.

Geach, Peter. 1957. *Mental Acts*. Routledge and Kegan Paul.

Gendler, Tamar Szabó. 2008a. "Alief and Belief." *Journal of Philosophy* 105:634–663.

Gendler, Tamar Szabó. 2008b. "Alief in Action (and Reaction)." *Mind and Language* 23:552–585.

Gibbard, Allan, and Hal R. Varian. 1978. "Economic Models." *Journal of Philosophy* 75:664–677.

Giere, Ronald N. 1988. "Explaining Science: A Cognitive Approach." *Philosophical Review* 100:653–656.

Gilbert, Margaret. 1989. *On Social Facts*. Routledge.

Glymour, Clark. 1980. "Why I Am Not a Bayesian." In "*Theory and Evidence*," was published in 1980 by Princeton University Press.

Goldstein, Simon. Forthcoming. "Omega Knowledge Matters." It won honorable mention for the 2021 Sanders prize in epistemology, and so is slated to be published in a future volume of *Oxford Studies in Epistemology*.

Good, I. J. 1966. "On the Principle of Total Evidence." *British Journal for the Philosophy of Science* 17:319–321.

Goodman, Jeremy, and Bernhard Salow. 2018. "Taking a Chance on KK." *Philosophical Studies* 175:183–196.

Goodman, Jeremy, and Bernhard Salow. Forthcoming. "Epistemology Normalized."

Greco, Daniel. 2013. "Probability and Prodigality." In *Oxford Studies in Epistemology*, Vol. IV, ed. Tamar Szabo Gendler and John Hawthorne, pp. 82–107. Oxford University Press.

Greco, Daniel. 2014a. "Could KK Be OK?" *Journal of Philosophy* 111:169–197.

Greco, Daniel. 2014b. "Iteration and Fragmentation." *Philosophy and Phenomenological Research* 88:656–673.

Greco, Daniel. 2014c. "A Puzzle about Epistemic Akrasia." *Philosophical Studies* 167:201–219.

Greco, Daniel. 2015. "Iteration Principles in Epistemology I: Arguments For." *Philosophy Compass* pp. 754–764.

Greco, Daniel. 2017. "Cognitive Mobile Homes." *Mind* 126:93–121.

Greco, Daniel. 2019. "Fragmentation and Higher-Order Evidence." In "*Higher-Order Evidence: New Essays*," is published by Oxford University Press. pp. 84–103.

Greco, Daniel. 2020. "Acting on Probabilistic Knowledge." *Res Philosophica* 97:109–117.

Greco, Daniel. Forthcoming. "Accessibilism without Consciousness."

Greco, John. 1990. "Internalism and Epistemically Responsible Belief." *Synthese* 85:245–277.

Hall, Ned. 1999. "How to Set a Surprise Exam." *Mind* 108:647–703.

Halpern, Joseph Y. 2003. *Reasoning about Uncertainty*. MIT Press.

Harman, Gilbert. 1986. *Change in View: Principles of Reasoning*. MIT Press.

Harman, Gilbert. 2001. "General Foundations versus Rational Insight." *Philosophy and Phenomenological Research* 63:657–663.

Hawthorne, John. 2004. *Knowledge and Lotteries*. Oxford University Press.

Hawthorne, John, and Maria Lasonen-Aarnio. 2009. "Knowledge and Objective Chance." In *Williamson on Knowledge*, ed. Patrick Greenough and Duncan Pritchard, pp. 92–108. Oxford University Press.

Hawthorne, John, and Ofra Magidor. 2009. "Assertion, Context, and Epistemic Accessibility." *Mind* 118:377–397.

Heal, Jane. 1978. "Common Knowledge." *Philosophical Quarterly* 28:116–131.

Heal, Jane. 1997. "Radical Interpretation." In *A Companion to the Philosophy of Language*, ed. Bob Hale, Crispin Wright, and Alexander Miller, pp. 175–196. Blackwell.

Heath, Joel P., H. Grant Gilchrist, and Ronald C. Ydenberg. 2010. "Interactions between Rate Processes with Different Timescales Explain Counterintuitive Foraging Patterns of Arctic Wintering Eiders." *Proceedings of the Royal Society B* 277:3179–3186.

Heath, Joseph. 2015. "Methodological Individualism." In "Stanford Enyclopedia of Philosophy," https://plato.stanford.edu/entries/methodological-individualism/.

Hedden, Brian. 2015. "A Defense of Objectivism about Evidential Support." *Canadian Journal of Philosophy* 45:716–743.

Hedden, Brian, and Mark Colyvan. 2019. "Legal Probabilism: A Qualified Defence." *Journal of Political Philosophy* 27:448–468.

Hempel, Carl G. 1945. "Studies in the Logic of Confirmation (II.)." *Mind* 54:97–121.

Hintikka, Jaakko. 1962. *Knowledge and Belief: An Introduction to the Logic of the Two Notions*. Cornell University Press.

Hirsch, Eli, and Matthias Jenny. 2019. "A Note on Safety and Iterated Knowledge." *Grazer Philosophische Studien* 96:244–254.

Holton, Richard. 2014. "Intention as a Model for Belief." In *Rational and Social Agency: Essays on the Philosophy of Michael Bratman*, ed. Manuel Vargas and Gideon Yaffe, pp. 12–37. Oxford University Press.

Jackson, Elizabeth. 2019. "How Belief-Credence Dualism Explains Away Pragmatic Encroachment." *Philosophical Quarterly* 69:511–533.

Jackson, Elizabeth G. 2020. "The Relationship between Belief and Credence." *Philosophy Compass* 15:1–13.

Jeffrey, Richard. 1965. *The Logic of Decision*. University of Chicago Press.

Jeffrey, Richard. 1970. "Dracula Meets Wolfman: Acceptance vs. Partial Belief." In *Induction, Acceptance, and Rational Belief*, ed. Marshal Swain, pp. 157–183 Reidel.

Johnson, Kent. 2015. "Notational Variants and Invariance in Linguistics." *Mind and Language* 30:162–186.

Johnson, Samuel, Avri Bilovich, and David Tuckett. 2022. "Conviction Narrative Theory: A Theory of Choice under Radical Uncertainty." *Behavioral and Brain Sciences*.

Kahneman, Daniel, and Amos Tversky. 1979. "Prospect Theory: An Analysis of Decision under Risk." *Econometrica* 47:263–292.

Kamm, Frances. 2007. *Intricate Ethics: Rights, Responsibilities, and Permissible Harm*. Oxford University Press.

Kaplan, David. 1977/1989. "Demonstratives." In *Themes from Kaplan*, ed. Joseph Almog, John Perry, and Howard Wettstein, pp. 481–564. Oxford University Press.

Kellert, Stephen H., Helen E. Longino, and C. Kenneth Waters. 2006. "The Pluralist Stance." In *Scientific Pluralism*, ed. Stephen H. Kellert, Helen Longino, and C. Kenneth Waters, pp. vii–xxviii. University of Minnesota Press.

Khalifa, Kareem. 2017. *Understanding, Explanation, and Scientific Knowledge*. Cambridge University Press.

Kim, Brian, and Anubav Vasudevan. 2017. "How to Expect a Surprising Exam." *Synthese* 194:3101–3133.

Kitcher, Philip. 1990. "The Division of Cognitive Labor." *Journal of Philosophy* 87:5–22.

Klemperer, Paul. 2004. *Auctions: Theory and Practice*. Princeton University Press.

Kotzen, Matthew. 2019. "A Formal Account of Epistemic Defeat." In *Themes from Klein*, ed. Cherie Braden, Rodrigo Borges, and Branden Fitelson, pp. 213–234. Springer Verlag.

Kriegel, Uriah, and Terrence Horgan. 2013. "The Phenomenal Intentionality Research Program." In *Phenomenal Intentionality*, ed. Uriah Kriegel, pp. 1–26. Oxford University Press.

Kuhn, Thomas S. 1962. *The Structure of Scientific Revolutions*. University of Chicago Press.

Lackey, Jennifer. 2007. "Norms of Assertion." *Noûs* 41:594–626.

Laplace, Pierre Simon. 1951. *A Philosophical Essay on Probabilities*. Dover Publications.

Lederman, Harvey. 2018a. "Two Paradoxes of Common Knowledge: Coordinated Attack and Electronic Mail." *Noûs* 52:921–945.

Lederman, Harvey. 2018b. "Uncommon Knowledge." *Mind* 127:1069–1105.

Leitgeb, Hannes. 2014. "The Stability Theory of Belief." *Philosophical Review* 123: 131–171.

Levins, Richard. 1966. "The Strategy of Model-Building in Population Biology." *American Scientist* 54:421–431.

Levitt, Steven, and Stephen Dubner. 2005. *Freakonomics*. William Morrow.

Lewis, David. 1974. "Radical Interpretation." *Synthese* 27:331–344.

Lewis, David. 1980. "Index, Context, and Content." In *Philosophy and Grammar*, ed. Stig Kanger and Sven Öhman, pp. 21–44. Reidel. Reprinted in Lewis 1998.

Lewis, David. 1982. "Logic for Equivocators." *Noûs* 16:431–441.

Lewis, David. 1983a. "New Work for a Theory of Universals." *Australasian Journal of Philosophy* 61:343–377. Reprinted in Lewis 1999.

Lewis, David. 1983b. "New Work for a Theory of Universals." *Australasian Journal of Philosophy* 61:343–377.

Lewis, David. 1986. *On the Plurality of Worlds*. Blackwell.

Lewis, David. 1996. "Elusive Knowledge." *Australasian Journal of Philosophy* 74:549–567. Reprinted in Lewis 1999.

Lewis, David. 1998. *Papers in Philosophical Logic*. Cambridge University Press.

Lewis, David. 1999. *Papers in Metaphysics and Epistemology*. Cambridge University Press.

Lewis, David. 2002/1969. *Convention: A Philosophical Study*. Blackwell.

Li, Ming, and Paul Vitányi. 2008. *Introduction to Kolmogorov Complexity and its Applications*. Springer.

MacFarlane, John. 2011. "Relativism and Knowledge Attributions." In *Routledge Companion to Epistemology*, ed. Bernecker Sven and Pritchard Duncan, pp. 536–544. Routledge.

Mandelbrot, Benoit. 1967. "How Long Is the Coast of Britain? Statistical Self-Similarity and Fractional Dimension?" *Science* 156:636–638.

Marcus, Ruth Barcan. 1980. "Moral Dilemmas and Consistency." *Journal of Philosophy* 77:121–136.

McAfee, R. Preston, and John McMillan. 1996. "Analyzing the Airwaves Auction." *Journal of Economic Perspectives* 10:159–175.

McGrew, Lydia M., and Timothy J. McGrew. 1997. "Level Connections in Epistemology." *American Philosophical Quarterly* 34:85–94.

McMullin, Ernan. 1985. "Galilean Idealization." *Studies in History and Philosophy of Science Part A* 16:247.

Meacham, Christopher. 2016. "Ur-Priors, Conditionalization, and Ur-Prior Conditionalization." *Ergo* 60:793–812.

Mills, Charles W. 2005. "'Ideal Theory' as Ideology." *Hypatia* 20:165–184.

Mitchell, Sandra D. 2003. *Biological Complexity and Integrative Pluralism*. Cambridge University Press.

Moon, Andrew, and Elizabeth Jackson. 2020. "Credence: A Belief-First Approach." *Canadian Journal of Philosophy* 50:652–669.

Morris, Stephen. 1995. "The Common Prior Assumption in Economic Theory." *Economics and Philosophy* 11:227–253.

Moss, Sarah. 2018. *Probabilistic Knowledge*. Oxford University Press.

Moss, Sarah. 2019. "Full Belief and Loose Speech." *Philosophical Review* 128:255–291.

Muth, John. 1961. "Rational Expectations and the Theory of Price Movements." *Econometrica* 29:315–335.

Myrvold, Wayne C. 2016. "Probabilities in Statistical Mechanics." In *The Oxford Handbook of Probability and Philosophy*, ed. Christopher Hitchcock and Alan Hájek, pp. 573–600. Oxford University Press.

Nagel, Jennifer. 2010. "Epistemic Anxiety and Adaptive Invariantism." *Philosophical Perspectives* 24:407–435.

Nolan, Daniel. 2014. "Hyperintensional Metaphysics." *Philosophical Studies* 171:149–160.

Norby, Aaron. 2015. "Uncertainty without All the Doubt." *Mind and Language* 30:70–94.

North, Jill. 2009. "The Structure of Physics: A Case Study." *Journal of Philosophy* 106:57–88.

Novaes, Catarina Dutilh, and Erich Reck. 2017. "Carnapian Explication, Formalisms as Cognitive Tools, and the Paradox of Adequate Formalization." *Synthese* 194: 195–215.

O'Connor, Cailin, and James Owen Weatherall. 2019. *The Misinformation Age: How False Beliefs Spread*. Yale University Press.

Okasha, Samir. 2001. "What Did Hume Really Show about Induction?" *Philosophical Quarterly* 51:307–327.

Okasha, Samir. 2005. "Bayesianism and the Traditional Problem of Induction." *Croatian Journal of Philosophy* 5:181–194.

Oppenheim, Paul, and Hilary Putnam. 1958. "Unity of Science as a Working Hypothesis." *Minnesota Studies in the Philosophy of Science* 2:3–36.

Parikh, Rohit. 2008. "Sentences, Belief, and Logical Omniscience, or What Does Deduction Tell Us?" *Review of Symbolic Logic* 1:459–476.

Pettigrew, Richard. 2021. "On the Expected Utility Objection to the Dutch Book Argument for Probabilism." *Noûs* 55:23–38.

Phillips, Jonathan, and Joshua Knobe. 2018. "The Psychological Representation of Modality." *Mind & Language* 33:65–94.

Pittard, John. 2019. *Disagreement, Deference, and Religious Commitment*. Oxford University Press.

Pollock, John L. 1987. "Defeasible Reasoning." *Cognitive Science* 11:481–518.

Posner, Richard. 1972. *Economic Analysis of Law*. Little Brown.

Priest, Graham. 2001. "Paraconsistent Belief Revision." *Theoria* 67:214–228.

Pryor, James. 2004. "What's Wrong with Moore's Argument." *Philosophical Issues* 14:349–378.

Pryor, James. 2013. "Problems for Credulism." In *Seemings and Justification: New Essays on Dogmatism and Phenomenal Conservatism*, ed. Chris Tucker, pp. 89–134. Oxford University Press.

Quine, W. V. 1953. "On a So-Called Paradox." *Mind* 62:65–67.

Quine, Willard Van Orman. 1960. *Word and Object*. MIT Press.

Quine, W. V. O., and J. S. Ullian. 1978. *The Web of Belief*. McGraw-Hill.

Rawls, John. 1971. *Theory of Justice*. Belknap Press.

Rayo, Agustín. 2008. "Vague Representation." *Mind* 117:329–373.

Rayo, Agustín. 2013a. *The Construction of Logical Space*. Oxford University Press.

Rayo, Agustín. 2013b. "A Plea for Semantic Localism." *Noûs* 47:647–679.

Rheinberger, Hans-Jorg, and Staffan Muller-Wille. 2015. "Gene." In "Stanford Enyclopedia of Philosophy," https://plato.stanford.edu/entries/impossible-worlds/.

Rice, Collin. 2018. "Idealized Models, Holistic Distortions, and Universality." *Synthese* 195:2795–2819.

Rice, Collin. 2019. "Models Don't Decompose That Way: A Holistic View of Idealized Models." *British Journal for the Philosophy of Science* 70:179–208.

Richardson, Lewis Fry. 1993. *Collected Papers of Lewis Fry Richardson*. Cambridge University Press.

Rinard, Susanna. 2014. "A New Bayesian Solution to the Paradox of the Ravens." *Philosophy of Science* 81:81–100.

Rodrik, Dani. 2015. *Economics Rules*. Oxford University Press.

Romero, Felipe. 2017. "Novelty versus Replicability: Virtues and Vices in the Reward System of Science." *Philosophy of Science* 84:1031–1043.

Ross, Jacob, and Mark Schroeder. 2014. "Belief, Credence, and Pragmatic Encroachment." *Philosophy and Phenomenological Research* 88:259–288.

Ross, Lewis. 2021. "Rehabilitating Statistical Evidence." *Philosophy and Phenomenological Research* 102:3–23.

Rubinstein, Ariel. 1989. "The Electronic Mail Game: Strategic Behavior under 'Almost Common Knowledge'." *American Economic Review* 79:385–391.

Rubinstein, Ariel. 2005. *Lecture Notes in Microeconomic Theory*. Princeton University Press.

Sainsbury, R. M. 1995. "Why the World Cannot Be Vague." *Southern Journal of Philosophy* 33:63–81.

Salow, Bernhard. 2017. "The Externalist's Guide to Fishing for Compliments." *Mind* 127:691–728.

Salow, Bernhard, and Arif Ahmed. 2019. "Don't Look Now." *British Journal for the Philosophy of Science* 70:327–350.

Samuelson, Larry. 2004. "Modeling Knowledge in Economic Analysis." *Journal of Economic Literature* 42:367–403.

Saul, Jennifer. 2013. "Scepticism and Implicit Bias." *Disputatio* 5:243–263.

Savage, Leonard J. 1954. *The Foundations of Statistics*. Wiley Publications in Statistics.

Schaffer, Jonathan. 2010. "The Debasing Demon." *Analysis* 70:228–237.

Schiffer, Stephen. 1972. *Meaning*. Clarendon Press.

Schoenfield, Miriam. 2016. "Moral Vagueness Is Ontic Vagueness." *Ethics* 126:257–282.

Schulz, Moritz. 2017. "Decisions and Higher-Order Knowledge." *Noûs* 51:463–483.

Schwarz, Wolfgang. 2018. "Imaginary Foundations." *Ergo: An Open Access Journal of Philosophy* 5:764–789.

Schwarz, Wolfgang. 2021. "Objects of Choice." *Mind* 130:165–197.

Schwitzgebel, Eric. 2001. "In-Between Believing." *Philosophical Quarterly* 51:76–82.

Schwitzgebel, Eric. 2010. "Acting Contrary to Our Professed Beliefs or the Gulf between Occurrent Judgment and Dispositional Belief." *Pacific Philosophical Quarterly* 91:531–553.

Schwitzgebel, Eric. 2011. *Perplexities of Consciousness*. MIT Press.

Sellars, Wilfrid. 1956. "Empiricism and the Philosophy of Mind." In *Minnesota Studies in the Philosophy of Science*, ed. Herbert Feigl and Michael Scriven, Vol. I, pp. 253–329. University of Minnesota Press.

Shannon, C. E. 1948. "A Mathematical Theory of Communication." *Bell Systems Technical Journal* 27:379–423.

Shoemaker, Sydney. 1994. "Lecture II: The Broad Perceptual Model—Self Knowledge and 'Inner Sense'." *Philosophy and Phenomenological Research* 54:271–290.

Sider, Theodore. 2011. *Writing the Book of the World*. Oxford University Press.

Singer, Daniel J., Aaron Bramson, Patrick Grim, Bennett Holman, Jiin Jung, Karen Kovaka, Anika Ranginani, and William J. Berger. 2019. "Rational Social and Political Polarization." *Philosophical Studies* 176:2243–2267.

Sklar, Lawrence. 1993. *Physics and Chance*. Cambridge University Press.

Skyrms, Brian. 1980. "Higher Order Degrees of Belief." In *Prospects for Pragmatism*, ed. D. H. Mellor, pp. 109–137. Cambridge University Press.

Skyrms, Brian. 2010. *Signals: Evolution, Learning, and Information*. Oxford University Press.

Smith, Holly M. 1988. "Making Moral Decisions." *Noûs* 22:89–108.

Smith, Martin. 2016. *Between Probability and Certainty: What Justifies Belief*. Oxford University Press.

Smith, Martin. 2018. "When Does Evidence Suffice for Conviction?" *Mind* 127: 1193–1218.

Smithies, Declan. 2019a. *The Epistemic Role of Consciousness*. Oxford University Press.

Smithies, Declan. 2019b. *The Epistemic Role of Consciousness*. Oxford University Press.

Soames, Scott. 1987. "Direct Reference, Propositional Attitudes, and Semantic Content." *Philosophical Topics* 15:47–87.

Sosa, David. 2009. "Dubious Assertions." *Philosophical Studies* 146:269–272.

Srinivasan, Amia. 2015. "Normativity without Cartesian Privilege." *Philosophical Issues* 25:273–299.

Staffel, Julia. 2019. *Unsettled Thoughts: A Theory of Degrees of Rationality*. Oxford University Press.

Stalnaker, Robert. 1978. "Assertion." *Syntax and Semantics (New York Academic Press)* 9:315–332.

Stalnaker, Robert. 1984. *Inquiry*. MIT Press.

Stalnaker, Robert. 1991. "The Problem of Logical Omniscience, I." *Synthese* 89:425–440.

Stalnaker, Robert C. 1996. "On What Possible Worlds Could Not Be." In *Benacerraf and his Critics*, ed. S. Stich and A. Morton, pp. 103–119. Blackwell.

Stanley, Jason. 2005. *Knowledge and Practical Interests*. Oxford University Press.

Stine, Gail. 1976. "Skepticism, Relevant Alternatives, and Deductive Closure." *Philosophical Studies* 29:249–261.

Strawson, Galen. 1994. *Mental Reality*. MIT Press.

Street, Sharon. 2006. "A Darwinian Dilemma for Realist Theories of Value." *Philosophical Studies* 127:109–166.

Strevens, Michael. 2006. "The Role of the Matthew Effect in Science." *Studies in History and Philosophy of Science Part A* 37:159–170.

Swanson, Eric. 2016. "The Application of Constraint Semantics to the Language of Subjective Uncertainty." *Journal of Philosophical Logic* 45:121–146.

Swinburne, Richard. 2001. *Epistemic Justification*. Oxford University Press.

Teller, Paul. 2001. "The Twilight of the Perfect Model Model." *Erkenntnis* 55:393–415.

Thomson, Judith Jarvis. 1985. "The Trolley Problem." *Yale Law Journal* 94:1395–1415.

Thomson, Judith Jarvis. 2008. "Turning the Trolley." *Philosophy and Public Affairs* 36:359–374.

Titelbaum, Michael G. 2010. "Not Enough There There Evidence, Reasons, and Language Independence." *Philosophical Perspectives* 24:477–528.

Titelbaum, Michael G. 2012. *Quitting Certainties: A Bayesian Framework Modeling Degrees of Belief*. Oxford University Press.

Tribe, Lawrence. 1971. "Trial by Matehmatics: Precision and Ritual in the Legal Process." *Harvard Law Review* 84:1329–1393.

Uzquiano, Gabriel, and Agustin Rayo. 2006. *Absolute Generality*. Oxford University Press.

van Fraassen, C. 1984. "Belief and the Will." *Journal of Philosophy* 81:235–256.

Veit, Walter. 2020. "Model Pluralism." *Philosophy of the Social Sciences* 50:91–114.

Velleman, David. 1989. *Practical Reflection*, vol. 94. Princeton University Press.

Vickrey, William. 1963. "Pricing in Urban and Suburban Transit." *The American Economic Review* 452–465.

Vogel, Jonathan. 1990. "Are There Counterexamples to the Closure Principle?". In *Doubting: Contemporary Perspectives on Skepticism*, ed. Michael David Roth and Glenn Ross, pp. 12–29. Kluwer Academic Publishers.

Vogel, Jonathan. 1999. "The New Relevant Alternatives Theory." *Nous* 33:155–180.

Wallace, David. 2012. *The Emergent Multiverse: Quantum Theory According to the Everett Interpretation*. Oxford University Press.

Warmbrōd, Ken. 1981. "Counterfactuals and Substitution of Equivalent Antecedents." *Journal of Philosophical Logic* 10:267–289.

Weinberg, Stephen. 1994. *Dreams of a Final Theory*. Vintage Books.

Weisberg, Jonathan. 2009. "Commutativity or Holism? A Dilemma for Conditionalizers." *British Journal for the Philosophy of Science* 60:793–812.

Weisberg, Jonathan. 2015. "Updating, Undermining, and Independence." *British Journal for the Philosophy of Science* 66:121–159.

Weisberg, Jonathan. 2020. "Belief in Psyontology." *Philosophers' Imprint* 20:1–27.

Weisberg, Michael. 2013. *Simulation and Similarity: Using Models to Understand the World*. Oxford University Press.

Wells, Gary. 1992. "Naked Statistical Evidence of Liability: Is Subjective Probability Enough?" *Journal of Personality and Social Psychology*. 62:739–752.

Whittle, Bruno. 2020. "Ontological Pluralism and Notational Variance." *Oxford Studies in Metaphysics* 12:58–72.

Williamson, Timothy. 1994. *Vagueness*. Routledge.

Williamson, Timothy. 2000. *Knowledge and its Limits*. Oxford University Press.

Williamson, Timothy. 2007. *The Philosophy of Philosophy*. Blackwell.

Williamson, Timothy. 2011. "Improbable Knowing." In *Evidentialism and its Discontents*, ed. Trent Dougherty, pp. 147–164. Oxford University Press.

Williamson, Timothy. 2014. "Very Improbable Knowing." *Erkenntnis* 79:971–999.

Williamson, Timothy. 2017. "Model-Building in Philosophy." In *Philosophy's Future*, ed. Russell Blackford and Damien Broderick, pp. 159–171.

Williamson, Timothy. 2018. "Counterpossibles." *Topoi* 37:357–368.

Wittgenstein, Ludwig. 1921. *Tractatus Logico-Philosophicus*. Routledge, 1961 edition.

Wylie, Alison. 1999. "Rethinking Unity as a 'Working Hypothesis' for Philosophy: How Archaeologists Exploit the Disunities of Science." *Perspectives on Science* 7:293–317.

Yablo, Stephen. 1992. "Mental Causation." *Philosophical Review* 101:245–280.

Yalcin, Seth. 2018. "Semantics as Model-Based Science." In *The Science of Meaning: Essays on the Metatheory of Natural Language Semantics*, ed. Derek Ball and Brian Rabern, pp. 334–360. Oxford University Press.

Yap, Audrey. 2014. "Idealization, Epistemic Logic, and Epistemology." *Synthese* 191:3351–3366.

Zimmerman, Aaron Z. 2018. *Belief: A Pragmatic Picture*. Oxford University Press.

Zollman, Kevin J. S. 2007. "The Communication Structure of Epistemic Communities." *Philosophy of Science* 74:574–587.

Zollman, Kevin J. S. 2018. "The Credit Economy and the Economic Rationality of Science." *Journal of Philosophy* 115:5–33.

Index